Impact of UV Disinfection on Biological Stability

The mission of the Awwa Research Foundation (AwwaRF) is to advance the science of water to improve the quality of life. Funded primarily through annual subscription payments from over 1,000 utilities, consulting firms, and manufacturers in North America and abroad, AwwaRF sponsors research on all aspects of drinking water, including supply and resources, treatment, monitoring and analysis, distribution, management, and health effects.

From its headquarters in Denver, Colorado, the AwwaRF staff directs and supports the efforts of over 700 volunteers, who are the heart of the research program. These volunteers, serving on various boards and committees, use their expertise to select and monitor research studies to benefit the entire drinking water community.

Research findings are disseminated through a number of technology transfer activities, including research reports, conferences, videotape summaries, and periodicals.

Impact of UV Disinfection on Biological Stability

Prepared by:
Graham A. Gagnon, **Trevor S. Dykstra**, and **Kevin C. O'Leary**
Dalhousie University, Department of Civil Engineering
1360 Barrington Street, Halifax, Nova Scotia

Robert C. Andrews
University of Toronto, Department of Civil Engineering
35 St. George Street, Toronto, Ontario

Christian Chauret
Indiana University Kokomo, Biological and Physical Sciences
Kokomo, Indiana 46904-9003

and

Christian Volk
Aqua Plus Services, Muncie, Indiana 47304

Sponsored by:
Awwa Research Foundation
6666 West Quincy Avenue, Denver, CO 80235-3098

Published by:

Awwa Research Foundation
Advancing the Science of Water®

IWA Publishing

DISCLAIMER

Published by IWA Publishing, Alliance House, 12 Caxton Street, London SW1H 0QS, UK

Telephone: +44 (0) 20 7654 5500; Fax: +44 (0) 20 7654 5555; Email: publications@iwap.co.uk;
Web: **www.iwapublishing.com**

AwwaRF report number 90999F. Originally published by AwwaRF for its subscribers in 2003.
IWA Publishing version published 2004.

ISBN 1 84339 894X

CONTENTS

TABLES

FIGURES

FOREWORD

The Awwa Research Foundation is a nonprofit corporation that is dedicated to the implementation of a research effort to help utilities respond to regulatory requirements and traditional high-priority concerns of the industry. The research agenda is developed through a process of consultation with subscribers and drinking water professionals. Under the umbrella of a Strategic Research Plan, the Research Advisory Council prioritizes the suggested projects based upon current and future needs, applicability, and past work; the recommendations are forwarded to the Board of Trustees for final selection. The foundation also sponsors research project through the unsolicited proposal process; the Collaborative Research, Research Applications, and Tailored Collaboration programs; and various joint research efforts with organizations such as the U.S. Environmental Protection Agency, the U.S. Bureau of Reclamation, and the Association of California Water Agencies.

This publication is a result of one of these sponsored studies, and it is hoped that its findings will be applied in communities throughout the world. The following report servers not only as a means of communicating the results of the water industry's centralized research program but also as a tool to enlist the further support of the nonmember utilities and individuals.

Projects are managed closely from their inception to the final report by the foundation's staff and large cadre of volunteers who willingly contribute their time and expertise. The foundation serves a planning and management function and awards contracts to other institutions such as water utilities, universities, and engineering firms. The funding for this research effort comes primarily from the Subscription Program, through which water utilities subscribe to the research program and make an annual payment proportionate to the volume of water they deliver and consultants and manufacturers subscribe based on their annual billings. The program offers a cost-effective and fair method for funding research in the public interest.

A broad spectrum of water supply issues is addressed by the foundation's research agenda: resource, treatment and operations, distribution and storage, water quality and analysis, toxicology, economics, and management. The ultimate purpose of the coordinated effort is to assist water suppliers to provide the highest possible quality of water economically and reliably. The true benefits are realized when the results are implemented at the utility level. The foundation's trustees are pleased to offer this publication as a contribution toward that end.

Edmund G. Archuleta, P.E.
Chair, Board of Trustees
Awwa Research Foundation

James F. Manwaring, P.E.
Executive Director
Awwa Research Foundation

ACKNOWLEDGMENTS

The members of the project team are grateful for the financial support provided by the Awwa Research Foundation. In particular, we acknowledge the efforts of our project manager, Ms. Kathryn Martin. As well, we also acknowledge the insightful technical contributions made our project advisory committee: Dr. Jamal Awad (CH2M Hill); Dr. Warren Jones (Montana State University), Dr. Bill Keevil (University of Southampton, UK), and Mr. Dan Pedersen (City of Austin Water and Wastewater Utility).

The project team are also thankful to Suntec Environmental Ltd. for providing a bench-scale UV reactor to conduct the research. Finally, the team is very appreciative of the in-kind support provided by our utility partners, namely East Bay Municipal Utility District, Halifax Regional Water Commission and the City of Ottawa.

EXECUTIVE SUMMARY

Interest in UV disinfection has increased significantly in recent years, due its ability to inactivate *Cryptosporidium parvum*. Although UV disinfection shows significant promise for inactivation of *Cryptosporidium*, very few investigations have examined the impact of UV disinfection on biological stability in distribution systems. Biologically stable water is defined as drinking water that does not promote the growth of microorganims, and is typically achieved either through the removal of substrate and/or by providing a secondary disinfectant. Because of increasing pressure to maintain disinfectant residual throughout the distribution system for most North American utilities, this research compared free chlorine, chloramines and chlorine dioxide. In particular, a goal of the research was to determine if there were any interactions or synergistic effects between UV and the downstream chemical disinfectants.

RESEARCH OBJECTIVES

The overall objective of the research is to evaluate the effect that UV disinfection has on biofilm growth in drinking water distribution systems. The research was conducted at bench-scale and employed annular reactors. The following additional objectives were also considered:

1. Determine the potential synergistic effects between primary disinfection with UV and secondary disinfection with chlorine-based disinfectants in controlling bacterial regrowth.
2. Compare the effectiveness of chlorine dioxide, free chlorine, and monochloramine for limiting bacterial regrowth in the presence and absence of UV disinfection.
3. Characterize the chemical intermediates formed during UV/Cl_2 photocatalysis of natural organic matter.
4. Quantify the impact of UV to impact bacterial regrowth in treatment plants that pre-chlorinate.
5. Examine the potential impact of UV/Cl_2 treatment on the biodegradability of natural organic matter (NOM) for three drinking water utilities.

APPROACH

The research was conducted in three phases: Phase 1 provided a direct comparison of chlorine-based disinfectants following UV disinfection; Phase 2 examined the biodegradability of humic material that had been photocatalyzed with UV/Cl_2; and Phase 3 characterized the organic by-products produced following UV/Cl_2 catalysis of post-filtered water obtained from three water utilities. Phases 1 and 2 were conducted at bench-scale using annular reactors (ARs) to monitor bacterial regrowth. Tap water consisting of surface water having a low alkalinity (30 mg/L as $CaCO_3$) and low concentration of total organic carbon (1.5 mg/L) was passed through two granular activated carbon (GAC) filters to remove background residual chlorine and biodegradable organic matter. For most of the experimental trials the GAC filtrate was pumped into a simulated UV disinfection clearwell and then subsequently pumped into the ARs. Along with the UV disinfected water, chlorine dioxide, free chlorine or monochloramine was dosed into the ARs to achieve effluent residual concentrations that are representative of typical full-scale applications. Phase 3 was conducted at bench-scale with a batch biodegradation study to monitor bacterial growth, by-product production and organic degradation. Post-filtered water was collected from three water utilities and BDOC determination was competed. Each organic source was treated with UV and a combination of UV and chlorine and transferred into eight flasks for analysis. HPCs, TOC, DOC, THMs, HAAs, aldehydes and carboxylic acids were determined over a 25-day period.

Prior to AR experimentation a model was developed to determine the effect of flow dynamics on the performance of potassium ferrioxalate as a chemical actinometer. Ultraviolet fluence was evaluated at several flow rates and calibrated using a bioassay with MS 2 bacteriophages. MS 2 was used to obtain a dose response curve for the reactor. A goal of at least 2-log inactivation of MS 2 was chosen since MS 2 is known to be a UV-resistant organism. The final objective was to quantify the dispersion occurring at flow rates investigated for fluence determination. Upon completion of AR experimentation the ability of a medium pressure UV lamp, operated at 45 mJ/cm^2, to inactivate MS 2 after a quartz tube has been fouled with organic rich source water, with and without pre-chlorination was determined. The inactivation of MS 2 was determined in the fouled and clean tube, in the presence and absence of organic rich water.

CONCLUSIONS

Impact of Ultraviolet Disinfection

- Disinfection with UV light was shown to enhance the potential for microbial control in drinking water distribution systems.
- When used without a secondary disinfectant UV alone was unable to control the proliferation of suspended or biofilm bacteria.
- Practically, these results would indicate that a dual-barrier disinfection system should be considered for pathogen protection at the plant and distribution system.

Impact of Secondary Disinfection

- Chlorine dioxide performed better than free chlorine on an equivalent Ct basis for the inactivation of suspended bacteria, whether or not UV disinfection was employed. Biofilm bacteria reacted similarly regardless of the disinfectant type or dose.
- Overall, these results indicate that the presence of secondary disinfectants within the distribution system is a crucial component required for microbial control.

Impact of Pre-chlorination

- Chlorination applied prior to UV disinfection, in the presence of NOM, acted to inhibit the development of biofilms and control the presence of bacteria in the bulk water.
- Pre-chlorination was able to limit the presence of bacteria to the same extent as was evident when secondary disinfection was applied when pre-chlorination was not used.
- The presence of chlorine during UV disinfection was also able to control the amount organic material deposited on the quartz tube as well as on the internal surfaces of the reactors.

- Overall, these findings indicate that the use of chlorination prior to UV disinfection improves microbial control in distribution systems, enhances the effectiveness of secondary disinfectants and potentially improves the efficiency of UV systems.

Role of NOM on UV disinfection

- The presence of NOM resulted in a similar suspended bacteria population as when NOM was not used. Conversely, NOM was shown to support a biofilm population that was 1-log higher.
- Reductions in heterotrophic bacteria (suspended and biofilm) were less in the presence of NOM than when the reactors were supplemented with BOM.
- NOM was shown to form deposits on the quartz tube and on the inner surfaces of the reactor.
- Oxidation of NOM was shown to increase bioavailability and the production of organic by-products, particularly HAAs and aldehydes.
- Overall, these results suggest that the amount of NOM in the source water entering the plant should be limited in order adequately irradiate the water and control biological growth.

Recommendations

1. Calibration of UV disinfection systems is important to ensure adequate disinfection. Calibration of UV lamps with collimated beams is unable to properly account for lamp fouling, and the water quality and hydraulic conditions that exist within reactors at the time of treatment. Calibration of UV disinfection system should be completed on site with water quality and hydraulic conditions that will exist during operation.
2. In areas where advanced water treatment exists, the addition of UV disinfection would enhance existing treatment barriers (e.g., filtration) for microbiological control. Following filtration the water is of very high quality, low turbidity and low colour; consequently, substantial UV disinfection would be possible with a minimal amount of quartz sleeve fouling. If UV light is used prior to filtration the water is of lower quality and depending of water quality, turbidity and colour, microbial inactivation can be reduced and quartz sleeve fouling may become an issue.
3. In areas where water treatment is nonexistent or minimal the addition of UV disinfection could be advantageous. The addition of UV light would provide a constant barrier against potentially pathogenic bacteria that may arise in the future. Although UV disinfection may offer an initial microbiological barrier it cannot control the regrowth of biofilm bacteria and secondary disinfection should be employed, particularly in locations where it was used in the past.

Future Research

Ultraviolet disinfection was shown to enhance microbial control when secondary disinfection was employed. Future research is required to further explore the impact of the various treatment methods investigated in this study. Additional pre-chlorination doses should be investigated to determine if lower concentrations are adequate to control regrowth events.

The effect of pre-chlorination and UV disinfection in the absence of humic material should also be investigated for comparison purposes.

The potential to produce UV disinfection by-products should be investigated further. Various by-products are known to form after oxidation with UV light. In this study, the concentration of HAAs and aldehydes both increased with increased oxidation. The impact of increased oxidation and oxidation type should be considered to determine what level of oxidation minimizes the concentration of DBPs and provides an adequate level of disinfection.

The relevance of using representative biological organic matter for nutrients in AR studies is of concern. In this study, humic material was shown to support a denser biofilm with higher total and heterotrophic bacterial cells than the synthetic cocktail. It is unclear at this time which of the two nutrient sources more closely models natural water conditions, and which should be used for further bench scale studies.

CHAPTER 1
PROJECT OVERVIEW

INTRODUCTION

Interest in ultraviolet (UV) radiation for primary disinfection of surface or ground water supplies is receiving increased interest from the water industry because several studies have indicated its ability to inactivate of *Cryptosporidium parvum oocysts*. In addition, proposed regulations concerning chlorinated disinfection by-products have resulted in a need to critically evaluate the application of UV disinfection in drinking water treatment. In particular, this project focuses on the effect of UV disinfection on bacterial regrowth in drinking water distribution systems.

UV is a physical disinfectant that leaves no residual protection. Thus, utilities that may consider the use of UV technologies generally focus on its ability to act as a primary disinfectant for inactivation of *Giardia* and *Cryptosporidium*. For these utilities, a secondary chlorine-based disinfectant (i.e., chlorine dioxide, chlorine, chloramines) will still be required to provide a final treatment barrier to protect consumers from acute (e.g., cross-connections) and chronic (e.g., growth of injured pathogenic microorganisms) microbiological problems in the distribution system. To date, very few studies have compared the effectiveness of UV/free chlorine and UV/monochloramine for controlling biofilm growth. This project compares these two chemical disinfectants in conjunction with UV and is the first known study to examine the effectiveness of UV and chlorine dioxide for biofilm control.

One concern with the application of UV technologies is the formation of chemical by-products, which may form during disinfection. To date, research has shown that the formation of chlorinated disinfectant by-products, such as trihalomethanes and haloacetic acids, is relatively minor under typical disinfection dosages. However, UV disinfection has been shown to photocatalyze hypochlorous acid (HOCl) to form free hydroxyl radicals (•OH). Hydroxyl radicals are highly unstable and react quickly with aromatic compounds, such as humic acids.

The reaction between •OH and humic material may lead to formation of carboxylic acids, aldehydes and other easily biodegradable organic matter. Thus there exists a potential (for utilities that add chlorine prior to UV disinfection) for a photocatalytic reaction to occur that involves UV, free chlorine and natural organic matter (NOM). The result of this reaction is the formation of biodegradable organic matter (BOM), which may serve as a nutrient source for biofilm growth in distribution systems. The characterization of BOM following UV/Cl_2 photocatalysis has not been reported in the literature. Although it is plausible that the by-products following UV/Cl2 photocatalysis consist of easily biodegradable material, there is a possibility that the by-products may not expose a distribution system to an increased risk for bacterial regrowth. This latter outcome would be of great significance to the water industry and would increase the confidence level associated with UV disinfection. It is anticipated that many water utilities will need to implement new disinfection strategies to comply with future water regulations.

The project evaluates the potential for UV/Cl_2 reactions to occur using synthetic drinking water in a model distribution system. In particular, the research employed annular reactors (ARs), which have been widely used in the water industry for evaluating biofilm growth under conditions representative of drinking water distribution systems. In addition, tests were performed on filtered water from three water utilities to characterize BOM following UV/Cl_2 photocatalysis.

RESEARCH OBJECTIVES

The overall objective of the research is to evaluate the effect of UV disinfection on biofilm growth in drinking water distribution systems. The research was conducted at bench-scale and employed annular reactors. The specific research objectives were

i. Determined the potential synergistic effects, which may control regrowth when applying UV at fluences required for *Cryptosporidium* inactivation, prior to the addition to chlorine based disinfectants.
ii. Compared the effectiveness of chlorine dioxide, free chlorine, and monochloramine for limiting bacterial regrowth in the presence and absence of UV disinfection.
iii. Characterized the chemical intermediates formed during UV/Cl_2 photocatalysis of natural organic matter.
iv. Quantified the impact of UV to impact bacterial regrowth in treatment plants that pre-chlorinate.
v. Examined the potential impact of UV/Cl_2 photocatalysis on biodegradability for three drinking water utilities.

The experiments were conducted in three phases to attain these objectives. The experimental phases were

- **Phase 1** provided a direct comparison of chlorine-based disinfectants following UV disinfection.
- **Phase 2** examined the biodegradability of humic material that has been photocatalyzed with UV/Cl_2.
- **Phase 3** characterized the organic by-products produced after UV/Cl_2 catalysis of post-filtered water from three water utilities.

REPORT ORGANIZATION

In general, the report is organized according to the project phases of this project. A general literature review is also provided for background and to provide a current state of the art for biofilm control and the impact UV disinfection has on distribution systems. In particular, the specific chapters address the following topics:

Chapter 2 provides an overview of biofilm control in distribution systems and UV disinfection and its potential to affect distribution system water quality.

Chapter 3 consists of general materials and methods that describe the techniques that are made use of throughout the experimental study. This chapter provides details of the model distribution and UV disinfection system as well as methods used throughout the study.

Chapter 4 describes the calibration of the UV disinfection system that was completed prior to the start of experimentation with the annular reactors. It describes the processes involved in determining the fluence supplied by the UV lamp as well as the formation of a model to estimate the fluence supplied over a reduced reaction length.

Chapter 5 compares the effectiveness of the secondary disinfectants as well as the potential for UV to control biofilm regrowth in distributions systems.

Chapter 6 evaluates the impact that UV disinfection and pre-chlorination have on NOM and biofilm growth.

Chapter 7 evaluates the impact that UV disinfection and pre-chlorination have on treated water samples from three water treatment plants.

Chapter 8 outlines the results from the fluence re-evaluation and the impact of organic fouling of the effectiveness of UV treatment.

Chapter 9 provides the overall project summary and recommendations to the water industry.

CHAPTER 2
LITERATURE REVIEW

REGROWTH IN DISTRIBUTION SYSTEMS

Regrowth events in drinking water distribution systems pose a significant problem to the drinking water industry. Bacterial regrowth has been described as the unexplained occurrence of coliforms in the distribution system (LeChevallier 1990, van der Wende et al. 1989, and Prévost et al. 1998), or as bacterial contamination outside the distribution system or multiplication of bacteria within the distribution system (Escobar and Randall 2001). In either description, regrowth is closely related to the availability of nutrients required for microbial growth, the ineffectiveness of secondary disinfectant residuals, and the accumulation of sediment or corrosion products. The presence of biofilms on distribution system pipe walls not only causes aesthetic issues, like taste and odour, but more importantly can also promote outbreaks of potentially pathogenic bacteria, which can lead to heath related problems (Escobar and Randall 2001, and Prévost et al. 1998). The concern towards bacterial regrowth in drinking water distribution systems stems from the presence and multiplication of potentially pathogenic bacteria like *Legionella pneumophila* (Escobar and Randall 2001). Van der Wende et al. (1989) found that the growth and detachment of biofilms account for most of the cells present in the bulk liquid phase. Lu et al. (1999) also reported that bacteria present in the bulk water originate from the biofilm present on the pipe walls of full-scale distribution systems.

In a study at a full-scale plant, LeChevallier et al. (1987) found that the treatment plant used for the study produced water with coliform levels well below established guidelines but found increased coliform levels as the water moved from the plant through the distribution system. Storey and Ashbolt (2001) investigated the impact of spiking virions into annular reactors (ARs) that had been previously established with biofilms. In that study it was found that virions have the potential to accumulate within distribution system biofilms and that problems could arise when the virions associated with biofilms detach into the bulk fluid. Coliforms are very often associated with soft pipe deposits and can also be the source and reservoir of pathogenic bacteria (Zacheus et al. 2001). Although coliforms were not found in the flowing water in that study, they could easily be released by disturbances in the water. Soft pipeline deposits were found to be the key site for microbial growth in distribution systems. The numbers of bacteria on soft deposits were significantly higher than the number in running water. Furthermore, Zacheus et al. (2001) reported that one-year after the mechanical cleaning of a section of a full-scale distribution system, the viable microbial population was almost as high as it had been previously.

DISINFECTION OF BIOFILMS

The control of distribution system water quality requires that the treatment process applies and maintains an effective disinfection concentration and removes the biodegradable organic matter in order to limit the bacterial density in finished water (Mathieu et al. 1993). Norton and LeChevallier (2000) found that once pipe surfaces are colonized, the protection levels against secondary disinfectants and growth rates are sufficient to maintain a substantial level of gram-negative cells in the biofilm despite being exposed to chlorinated waters containing gram-positive bacteria. Bacterial flux also has an impact on the development of a biofilm; limitation of the flux of cells entering the network will decrease the density of biofilm (Mathieu et al. 1993). Lu et al. (1999) identified that chlorine is consumed in the bulk liquid phase by iron, organic compounds, and by the distribution system pipe walls (iron and biofilm). A chlorine concentration of 3-5 mg/L was able to reduce the biofilm heterotrophic bacteria by 3-logs. A 2-log reduction was observed when 1 mg/L of chloramines were used (LeChevallier 1990). When an influent free chlorine concentration of 0.8 mg/L was used, bacteria only grew in the protective environment of the biofilm and not in the water phase (van der Wende et al. 1989). Suspended bacteria did not appear in the distribution system until the free chlorine disappeared despite the presence of a biofilm (Cipparone et al. 1997). The use of chlorine may lead to the selection of a limited number of microbial species in the early stages of biofilm accumulation and reduce the accumulation of biofilm and influences its distribution (van der Wende et al. 1989).

Biofilm growth and detachment may be even more dominant in water distribution systems with (low) chlorine concentrations than in chlorine-free systems. Despite the use of continuous disinfection, biofilms rapidly develop within distribution systems (Norton and LeChevallier 2000). Population shifts during chlorination indicate that chlorine disinfection plays a substantial role in determining the composition of suspended microbial populations (Norton and LeChevallier 2000). Camper et al. (1996) found that chlorine is unable to control regrowth events because the so-called microcolonies are less susceptible to disinfection. Another factor may be that the concentration of chlorine at surfaces in flowing water is considerably less than in the fluid (Pedersen 1990).

ORGANIC MATTER IN DISTRIBUTION SYSTEMS

The three nutrients that are considered most important to the proliferation of microorganisms in drinking water are carbon, nitrogen, and phosphorous. For heterotrophic bacteria, the general requirements for carbon, nitrogen, and phosphorous are in a ratio of about 100:10:1 (Chandy and Angles 2001, Percival et al. 2000, Camper 1996). Biodegradable organic matter (BOM) is considered to be the limiting nutrient for regrowth in drinking water distribution systems (LeChevallier 1990) and to be the primary energy and carbon source in drinking water (Rittmann and Huck 1989). There is little evidence to support the theory that there is a similar correlation for both nitrogen and phosphorous. Studies that have been completed suggest that such a correlation does not exist (Percival et al. 2000), although phosphorous at low concentrations may restrict microbial growth. Miettinen et al. (1997) reported that their investigation into the effect phosphorous has on microbial growth resulted in a positive correlation. Phosphorous concentrations are not required to be high to stimulate bacterial growth — the addition of 1 mg of PO_4-P L^{-1} is sufficient. Bacterial growth continues to increase up to a phosphate concentration of 10 μg of PO_4-P L^{-1}. This group also reported that other organic nutrients did not significantly affect microbial growth, this included nitrogen and potassium. As a result, the presence of carbon and the ability of microorganisms to use the available carbon are of prime importance.

Biological organic matter (BOM) can lead to bacterial regrowth in drinking water distribution systems, which in turn can lead to the deterioration of water quality, accelerate corrosion rates of pipes, and has the potential to increase the incidence of bacteriological disease (Escobar and Randall 2001, and Prévost et al. 1998). The removal of organic material can help avoid the problems of taste and odour (Agbekodo et al. 1996). In a study completed by Escobar and Randall (2001), ozonating water oxidized the larger organic molecules, which increased the assimilable organic carbon (AOC) content in the distribution system by 200% over the short term and 100% over the long term. The AOC increased from 70 μg/L acetate-C to approximately 200 μg/L acetate-C. This significant increase in the amount of carbon available to the microbes present resulted in a three-fold increase in bacterial counts over the short term and a two-fold increase over the long term. Zacheus et al. (2001) reports on a study completed by van der Kooij that states AOC levels below 10 μg acetate-C/L are biologically stable. LeChevallier et al. (1991) reports that regrowth is still possible at acetate-C concentration of 50 μg/L. However, Zacheus et al. (2001) reports that microbes in old deposits recirculate nutrients, which results in a decreased nutrient requirement from the water.

In a study completed by Reckhow et al. (1996) DOC levels were shown to decrease as raw water was treated and traveled through the distribution system, which shows that bacterial utilization occurs in the distribution system. As a result it is important to minimize the level of natural organic matter in water in order to reduce regrowth events (Agbekodo et al. 1996, Lu et al. 1999, and van der Wende et al. 1989). In a full-scale study Agbekodo et al. (1996) found that nanofiltration, as a means of controlling organic loading, produced a permeate of high quality, which is low in DOC, BDOC and as a result would likely be low in DBPs once chlorinated. In a similar membrane study conducted on a groundwater source Nobel et al. (1996) showed that a membrane filtration treatment train produced water that was biologically unstable and that conventional treatment (coagulation, sedimentation, and filtration) produced water that was more biologically stable.

Organic matter, both dissolved and particulate, and the nutrient loading rate contribute to the growth and activity of heterotrophic bacteria in a distribution system (Escobar and Randall 2001, and van der Wende et al. 1989). Gauthier et al. (1999) reported that suspended particles can generate at least two types of problems for distribution water quality: (1) they carry bacteria fixed to the surface, which protect them from disinfection and (2) they contribute to the formation of loose deposits, which exacerbates the biofilm problem. Organic matter is suspected to greatly influence the activity of bacteria inside loose deposits (Gauthier et al. 1999). LeChevallier (1990) reports that the presence of corrosion and sediment accumulation can protect attached bacterial growth from disinfection, and an increase in the amount of free chlorine used to control biofilm regrowth also increased the oxidation of the organics into more useable forms (Escobar and Randall 2001). Organic matter and inorganic matter can accumulate on the surface of drinking water pipelines (Zacheus et al. 2001). Camper et al. (1996) used both mild steel (MS) and polycarbonate (PC) coupons for experimentation. In one test they replaced a MS coupon with a PC coupon and found that it supported less heterotrophic bacteria than the MS but more than the reactors with PC coupons exclusively. This was thought to be the result of an accumulation of corrosion products on the clean PC coupons. Coupons in the MS reactors supported 10 times more HPC bacteria than the PC reactors. Humic material used as a nutrient source has also been shown to significantly increase effluent total cells and culturable cells even after chlorination was initiated (Ellis et al. 2000).

INTERACTION BETWEEN UV LIGHT AND ORGANIC MATTER

Treatment of water with ultraviolet light has increased in popularity in recent years. This has been the result of improvements in UV technology (Backlund 1992) and the

Disinfectants/Disinfection By-Product (D/DBP) Rule that limits the level of disinfectants used and by-products produced as a result of the humic material present in the water source. Chemical disinfectants play a large role in the production of DBPs like trihalomethanes (THMs) and haloacetic acids (HAAs). The main appeal of using UV light is the fact that hazardous by-products are not produced coupled with the substantial microbial inactivation.

Studies have also been conducted to determine the effect that UV radiation has on the humic material itself. Backlund (1992) reported that UV radiation affects the composition of dissolved organic material (DOM) in natural waters. Humic material in natural waters is a result of decomposition of organic matter, thus presenting an opportunity for the production of DBPs and interference with disinfection. Allard et al. (1994) reported that UV radiation is strongly absorbed by humic substances in aquatic systems. Humic substances when exposed to UV light can be affected in various ways. Ultraviolet radiation can reduce the TOC content, colour of humic rich water, and reduce molecular size (Gjessing and Källqvist 1991, Corin et al. 1996). The result is a change in chemical composition that produces new biological properties that has the potential to restrain microbial growth. Lund and Hongve (1994) report similar findings. Their study reported that water irradiated with UV containing humic substances creates new components, which can inhibit bacterial growth and reduce biofilm formation in the pipes of a pilot plant. Studies conducted on *Selenastrum capicornutum* resulted in growth inhibition lasting several weeks. The authors concluded that toxic components are produced during UV treatment. Backlund (1992) reported findings that suggest that UV does not reduce the threat of DBPs. It was stated that using UV irradiation as a pretreatment step cannot decrease substantial mutagenic activity formed during chlorination of humic waters. Interactions between UV and humics lead to the direct photochemical transformations of the humic material. These can then act as precursors for the production of reactive intermediates.

Previous studies have reported that particulate material present in the water can act to reduce the exposure of UV light to the receiving waters. Qualls et al. (1983) showed that bacterial adsorption to clays provides little protection but the interference of organic particles can serve to protect bacteria and viruses from intended disinfection. Humic substances are a part of natural waters and are the result of decomposed organic material. The presence of humic material in source water that is irradiated can be a limiting factor on the biocidal effect of UV light. Qualls et al. (1983) reported that organic particles can become a major limiting factor in disinfection. This limitation is attributed to humic substances being the most highly UV absorbing compounds in natural waters (Lund and Hongve 1994). Corin et al. (1996) explained that the absorption of UV by humic substances to be due to conjugated bonds along with the presence of free electron pairs on heteroatoms. Suspended particles in water can play two roles, they can absorb and scatter UV light and they can harbour bacteria to protect them from light (Qualls et al. 1983). Bitton et al. (1972) conducted a study to determine the effect that clay minerals and humic acid have on the survival of *Klebsiella aerogenes* under a lethal dose of UV irradiation. They found that clay minerals and humic acid provided some protection. This protection was a result of the light absorbance and light scattering properties of both clay and humic acid. The study also suggests a possible role of iron compounds in protecting microorganisms against UV radiation. Qualls et al. (1983) suggested that clays do little to inhibit UV disinfection reporting that clay tends to scatter rather than absorb UV light. In addition to the influence shielding or protection, Bukhari et al. (1999) found that the absorbance of the water can reduce the amount of UV delivered to the target organism.

INTERACTION BETWEEN UV LIGHT AND CHLORINE

Previous studies have shown that UV light at sufficiently high fluences can promote the photolysis of chlorine in water to generate •Cl and •OH radicals (Zheng et al. 1999, Nowell and Hoihné 1992a, Nowell and Hoihné 1992b). These radicals are very strong oxidants, which are capable of reacting with a variety of organic substances in water (Zheng et al. 1999). Results from the studies completed by Zheng et al. (1999) and Nowell and Hoihné (1992a, 1992b) have concluded that at UV fluences used for drinking water treatment the degradation of chlorine is very minimal.

DISINFECTION OPTIONS

Trussel (1999) described disinfection as the most important step in any water treatment process train. The process of disinfection is the destruction or inactivation of pathogenic microorganisms, including bacteria, amoebic cysts, algae, spores and viruses (Montgomery 1985). Disinfectants are generally divided into two groups: (1) chemical agents and (2) nonchemical agents. Chemical agents include chlorine, chlorine dioxide, chloramines, and ozone. Nonchemical agents include ultraviolet and gamma radiation. The function of a disinfectant can be broken into primary and secondary disinfection (Trussel 1999). Primary disinfection removes or inactivates microbial contaminants from raw water to produce microbiologically safe treated water. Secondary disinfection involves residual maintenance in the distribution system.

Ultraviolet Light

Ultraviolet light is an attractive disinfectant due to its ability to inactivate bacteria and viruses (Battigelli et al. 1993). Ultraviolet radiation attacks nucleic acids by photochemical dimerization of pyrimidine bases (Shaw et al. 2000). At the proper fluence and wavelength ultraviolet light can damage the DNA and destroy the microorganism's ability to reproduce. Treatment of drinking water with ultraviolet light has increased in popularity in recent years because of its potential to inactivate *Cryptosporidium parvum* and *Giardia lamblia* (Craik et al. 2000, Craik et al. 2001), with a very low potential for the formation of harmful disinfection by-products (Cotton et al. 2001). Oocysts are resistant to chlorine at concentrations used in water treatment (Mofidi et al. 2001), and can cause potentially serious gastrointestinal illnesses in humans (Craun et al. 1998). Previous studies have reported that particulate material present in the water can act to reduce the exposure of UV light to the receiving waters. Qualls et al. (1983) reports that bacterial adsorption to clays provides little protection but the interference of organic particles can serve to protect bacteria and viruses from intended disinfection.

For drinking water treatment applications the target UV fluence ranges from 30 to 80 mJ/cm^2, which represents a low fluence for bacterial inactivation and high fluence for viral disinfection (Linden and Mofidi 1999). Considerable bacterial inactivation can be achieved at a low fluence. A 2-log inactivation of *Giardia muris* cysts can be achieved at a fluence of 5 mJ/cm^2 using a medium pressure UV source (Craik et al. 2000). Substantial oocyst inactivation (3 to 4-log) can be achieved at fluences from 19 to 25 mJ/cm^2, using medium pressure UV lamps (Bukhari et al. 1999, and Craik et al. 2001). Battigelli et al. (1993) reports that fluences of 25 mJ/cm^2 resulted in a 3-log inactivation of most test viruses although considerable resistance of MS 2 to UV light was reported indicating that larger fluences are required to achieve extensive virus reduction. In studies completed by Cotton et al. (2001), UV disinfection can achieve adequate *Cryptosporidium* inactivation and is relatively inexpensive. Costs were 40-80% lower for a >2-log inactivation of *Cryptosporidium* than for a 1-2 log inactivation with ozone. While these studies have demonstrated the capabilities of UV disinfection in controlled experiments, the ability to quantify and control UV fluence under full-scale operation is essential.

Secondary Disinfection Options

Secondary disinfectants are used to provide a residual concentration in a distribution system. There are many characteristics of a good disinfectant, three important ones being (Droste 1997):

1. Effective kill of pathogenic microorganisms
2. Nontoxic to humans or domestic animals
3. Provides residual protection in drinking water

Secondary disinfectants are predominantly chemical agents such as chlorine and chloramines, both of which provide residual protection in distribution systems. Others include chlorine dioxide and ozone, the latter of which does not provide any residual protection. Ozone has the highest oxidation potential of the disinfectants listed followed by chlorine dioxide, chlorine and chloramines. When comparing the disinfectants on the basis of mutagenic material produced, chlorine is the highest followed by chloramines, chlorine dioxide and ozone.

In large part, North American water utilities utilize secondary disinfectants to control regrowth events and protect consumers from potential microbial contact. Approximately one-third of water utilities in the Netherlands do not apply secondary disinfectants due to concerns over disinfection by-products but instead concentrate on the removal of precursors of microbial growth (van der Kooij et al. 1999). It should be noted that microbial outbreaks have occurred as a result of inadequate or interrupted disinfection (Droste 1997).

Chlorine

Free chlorine has been applied to drinking water for microbial protection since the early 1900s. The stability of the chlorine residual is good in finished waters with low concentrations of natural organic matter (NOM) and in distribution systems composed of inert materials. However, it is well established that free chlorine can react with natural organic matter (NOM) to form harmful DBPs like trihalomethanes. The formation of DBPs will most likely limit the future use of chlorine (Trussel 1999) and has led the water industry to actively seek alternative disinfectants.

Chloramines

Chloramines result from reactions between chlorine and ammonia. The species of chloramines formed are a result of chlorine to ammonia ratio and pH (Montgomery 1985). The principle reactions for the formation of chloramines are shown in Equations 2.1 – 2.3. The formation of di- and tri-chloramines are not favoured at high pH.

$$NH_3(aq) + HOCl \leftrightarrows NH_2Cl + H_2O \tag{2.1}$$

$$NH_2Cl + HOCl \leftrightarrows NHCl_2 + H_2O \tag{2.2}$$

$$NHCl_2 + HOCl \leftrightarrows NCl_3 + H_2O \tag{2.3}$$

Monochloramine can be a better biocide for controlling biofilm growth (Norton and LeChevallier 1997), particularly in ferrous metal pipes (Trussel 1999). However, when considering inactivation of resistant microorganisms such as *Giardia* and *Cryptosporidium,* monochloramine has a lower disinfection potential than free chlorine and is typically not applied to provide primary disinfection. Monochoramine is also subject to nitrification in the distribution system, which can lead to residual destruction.

Chlorine Dioxide

Chlorine dioxide is a powerful oxidizing agent and has been used in the past for taste and odour removal (Droste 1997), but its application has been limited for disinfection purposes (Volk et al. 2002). Chlorine dioxide is also more effective than chlorine over a wider pH range and does not react with organic precursors and as a result does not produce a significant amount of DBPs (Volk et al. 2002), although there is some evidence that suggests that ClO_2 produces other reaction products including aldehydes, carboxylic acids and ketones (AWWA and ASCE 1998). A draw back to the use of ClO_2 for disinfection is its toxicity and that of the chlorite ion it produces as it oxidizes materials in the water (Trussel 1999).

CHAPTER 3
MATERIALS AND METHODS

INTRODUCTION

This chapter is intended to provide a general description of the experimental systems, and the materials and methods used for most of the experiments conducted in this research. Most of the procedures are similar to or adapted from standard water quality procedures. A novel method for calibrating the model UV system was also developed as part of this research. Because this method is relatively new, a separate chapter (Chapter 4) was prepared to highlight this technique. Other specific experimental designs or variations of experimental set-up are described in subsequent chapters where it is appropriate.

OVERVIEW OF BENCH-SCALE TREATMENT SYSTEM

A general schematic of the model drinking water distribution system used in this study is provided in Figure 3.1. Prior to entering the ultraviolet (UV) lamp and annular reactors (ARs), tap water is passed up flow into a granular activated carbon (GAC) filter to remove any free chlorine present and then passed into a biologically activated carbon (BAC) filter to reduce the background organic carbon concentration and supply the distribution system with an inoculum of microorganisms. The resulting water is collected in a clearwell and is subsequently pumped through a UV disinfection system and collected in another clearwell. From the UV clearwell the water is pumped into each of the annular reactor (ARs) which are augmented with a cocktail of carbon, nitrogen and phosphorous, and a disinfectant.

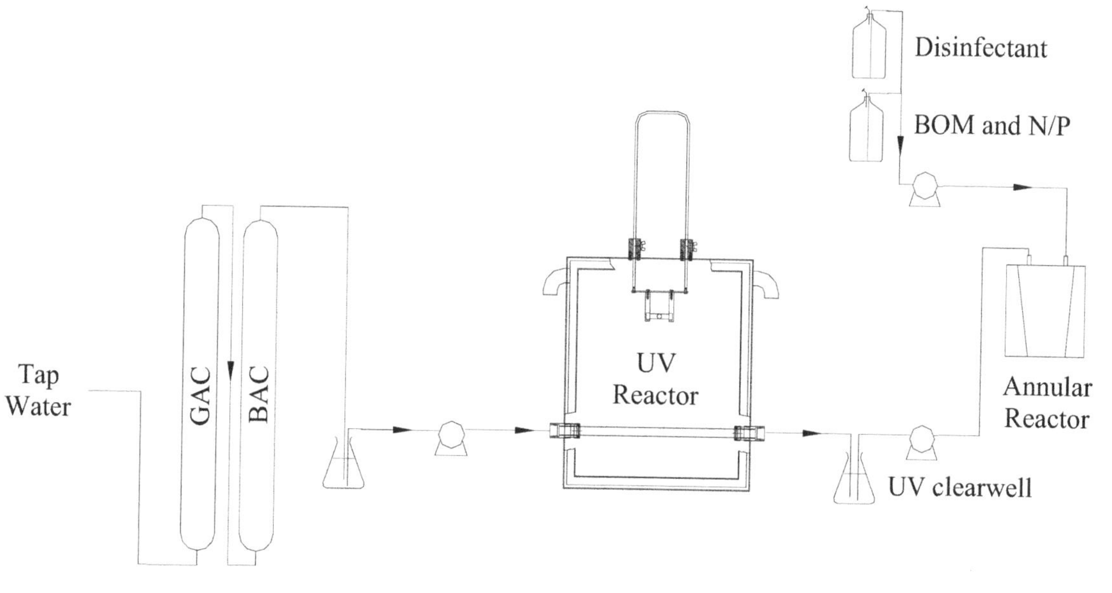

Figure 3.1 General schematic of the model drinking water distribution system

BENCH-SCALE TREATMENT SYSTEM

Raw Water Characteristics

Tap water supplied by the Halifax Regional Water Commission (HRWC) was used as the primary process water for the ARs. On average, the treated surface water has an alkalinity of 35 mg/L as $CaCO_3$, pH of 8.5, and total organic carbon (TOC) concentration of approximately 1.5 mg/L. The HRWC typically distributes water that has a free chlorine residual of approximately 0.6 mg/L.

GAC/BAC Filters

Tap water was passed up flow through a filter column containing fresh granular activated carbon (GAC) media to neutralize the free chlorine present (Figure 3.2). A second GAC filter operating biologically was used to remove any background natural organic matter (NOM) in the tap water. The biological activated carbon (BAC) media for the second GAC filter was obtained from biological filters at the Mannheim Water Treatment Plant in Waterloo, ON. The filtration columns (Kontes, Vineland, NJ) are operated up flow and are constructed of double-paned glass.

Ultraviolet Disinfection System

A medium pressure UV lamp and quartz tube is situated in a ventilated galvanized casing. A schematic of the UV lamp assembly (SUNTEC Environmental, Concord, ON) is provided in Figure 3.3. This configuration allows for UV fluence to be varied by positioning the UV lamp at various vertical heights above the quartz tube. A blower located at the bottom of the unit cools the lamp.

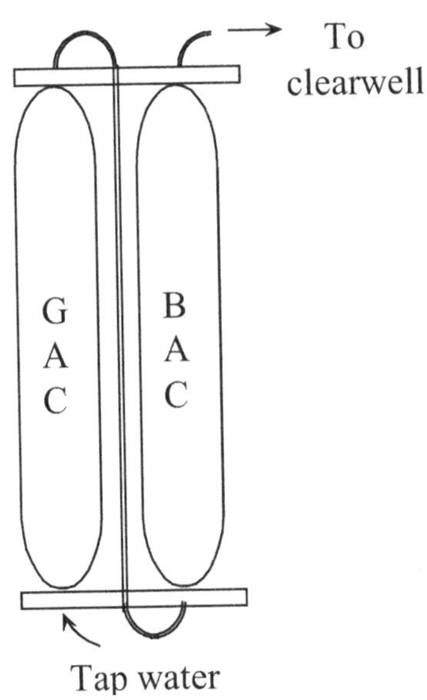

Figure 3.2 GAC/BAC filtration set-up

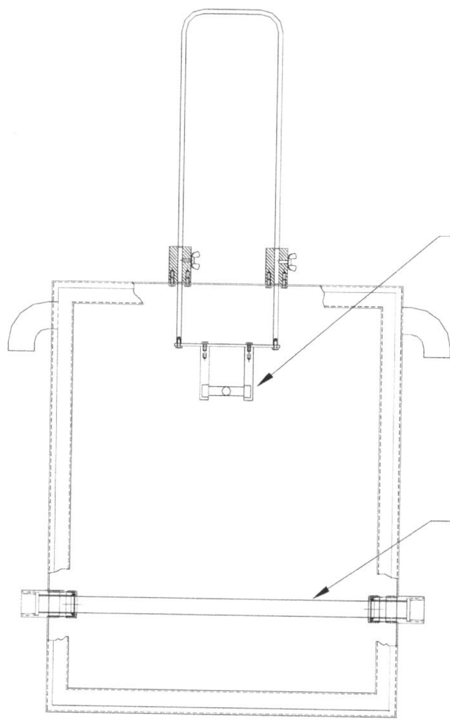

Figure 3.3 General schematic of the ultraviolet disinfection reactor (*Source*: Dykstra et al. [2002]. Reprinted from Hydraulic Calibration and Fluence Determination of Model UV Disinfection System, *ASCE – J. Enviro. Engrg.*, by permission. Copyright © 2002, ASCE.)

Water was pumped through the reactor with a variable speed modular peristaltic pump and master flex precision tubing (Saint-Gobain Performance Plastics, Wayne, NJ). A static in-line pipe mixer containing 6 mixing elements (Koflo Corporation, Cary, IL) was installed just prior to the inlet of the quartz tube to assist with mixing before application to the quartz tube. The mixer is constructed from stainless steel and is 2.54 cm in diameter and 22.9 cm in length. The calibration procedure to characterize hydraulic profile within the reactor tube and to determine the fluence of the model UV system is described in detail in Chapter 4.

Operation

Before use and between each experimental trial the quartz tube was thoroughly cleaned and the entire system (quartz tube and UV clearwell) was sterilized with 70% ethanol by continuous recirculation for a period of at least 24 hours. The lamp was also turned on 30 minutes prior to irradiation to ensure a uniform lamp output. Flow through the quartz tube was determined as described in Chapter 4 and set prior to each experimental trial.

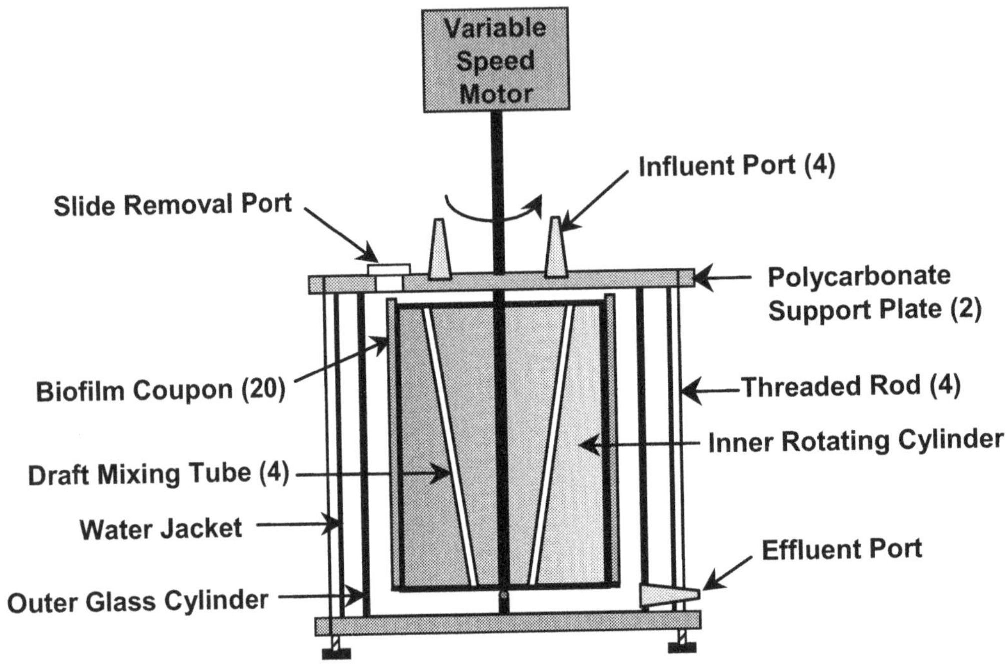

Figure 3.4 General schematic of an annular reactor

Annular Reactor

The annular reactors (ARs) (Model LJ, BioSurface Technologies Corporation, Bozeman, MT) are constructed of glass, polycarbonate and stainless steel, and can be manually cleaned and disinfected with ethanol (Figure 3.4). Each AR consists of a stationary outer glass cylinder and an inner drum that rotates via a variable speed motor located above the reactor. Water pulled through an annular gap by the rotation of the inner drum. Shear stress in the AR is controlled by the rotational speed of the inner drum and hydraulic retention time (HRT) is controlled independently by the volumetric flow rate. Biofilm accumulates on exposed surfaces and is sampled from 20 removable coupons, which are flush mounted to the surface of the inner drum. The inner drum contains draft tubes to ensure good vertical as well as horizontal mixing; a mixing time of less than 60 s has been reported (Gjaltema et al. 1994). The coupons used are manufactured from polycarbonate and can be easily sampled aseptically using a sterile hook.

The reactors are capable of receiving multiple influent lines, which allows for substantial flexibility in experimental design. In the reactor, water flows through the annulus that is created between the inner rotating cylinder, the glass cylinder and the support plates. The total working volume in the annular gap is approximately 950 mL.

Operation

Before use and between each experimental trial the ARs were thoroughly cleaned with soap and water. The entire system (ARs and influent/effluent lines) was sterilized with 70%

ethanol by continuous recirculation for a period of at least 24 hours. Flow through the ARs and from the cocktails and disinfectants were determined and set prior to experimentation.

Biostability experiments employed the use of four annular reactors. Each AR was maintained at a fixed rotational speed of 50-rpm, which is equivalent to a shear stress of 0.25 N/m^2 at the outer wall (Camper 1995). This shear stress approximately corresponds to a flow of 0.30 m/s (1 fps) in a 100-mm (4 in.) diameter smooth pipe and is similar to the shear conditions of other pilot- and bench-scale investigations (Camper 1996). The retention time for each AR was 1 hour to ensure that the growth of suspended cells was negligible in comparison to the growth rate of biofilm cells and biofilm detachment rate. Influent nutrient and disinfectant feed rates were determined based on stock concentration and the retention time within the reactor. Exposed surfaces (AR, tubing, feed bottles) were covered by an opaque material to reduce the potential of phototrophic growth in the bench-scale system. Experiments were conducted at a temperature of 20 ± 1°C.

Synthetic Waters

Organic Carbon, Nitrogen and Phosphorous Cocktails

Influent cocktails of organic carbon, nitrogen ($NaNO_3$) and phosphorous (K_2HPO_4 and KH_2PO_4) were prepared in sterile 4-L amber bottles. The cocktail solution of biodegradable organic material (BOM) was composed of ethyl alcohol, propionaldehyde, oxalate, pyruvate, and acetate to an organic carbon concentration of 100 μg C/L. These compounds were selected because they represent major classes of organic compounds found in drinking water (Camper 1996). They are also relatively non-reactive with disinfectants. Each compound was added in equimolar concentrations on the basis of oxygen demand. The cocktails were pumped into the ARs to obtain a molar C:N:P ratio of 100:20:5 and at a final concentration of 100 μg C/L.

Commercial Humic Acid

The commercial humic acid (Aldrich humic acid) was prepared to a TOC concentration of approximately 210 mg/L and diluted with biologically activated carbon (BAC) water to a concentration of 2.5 mg/L. Table 3.1 shows the elemental components of the humic acid on a percent-by-weight basis.

SAMPLE COLLECTION AND ANALYSIS

Microbial Samples

Bulk water samples for suspended microbial analysis were collected on a routine basis (weekly) at the effluent of the annular reactors as well as various locations prior to the inlet of the reactors (tap water, post BAC, post UV). Biofilm samples for fixed microbial analysis were also collected on a routine basis. Heterotrophic plate counts (HPC), acridine orange direct counts (AODC) and Live/dead cell enumerations were performed on collected samples.

Bulk water samples (10 mL) for HPC and Live/dead enumeration were collected in sterile 15-mL disposable, plastic tubes (Corning Inc., Acton, MA) containing 10-μL of 10% (w/v) sterile sodium thiosulfate. Bulk water samples (9 mL) for AODC were collected in sterile

Table 3.1

Elemental content of Aldrich humic acid

	C*	H	O	N	S	P	Total	Ash (%)
Aldrich Humic Acid, sodium salt	68.98	5.26	43.45	0.74	4.24	<0.05	122.7	31.21

Source: adapted from MacCarthy and Malcolm 1987.

* All results are expressed on a percent-by-weight, ash-free, and moisture-free basis.

15-mL disposable, plastic tubes (Corning Inc., Acton, MA) containing 1-mL of 5% (w/v) glutaraldehyde in PBS.

Polycarbonate coupons were removed, in sequence, aseptically from the annular reactor and transferred into 150-mL sterile test tube containing 25 mL phosphate buffered saline (PBS) and 25-µL of 10% (w/v) sterile sodium thiosulfate. The attached cells were immediately removed as described by Gagnon and Slawson (1999). The coupons and saline solution were transferred to a sterile stomacher bag (177 × 304 mm) and placed into the stomacher (Stomacher 400 Lab Blender, Seward, London, UK). The stomacher was operated at 230 rpm for 2 minutes to ensure adequate removal. The resulting bulk liquid was then transferred aseptically into sterile 15-mL disposable plastic tubes (Corning Inc., Acton, MA). The removed coupons were cleaned, sterilized and replaced to maintain constant volume and hydraulic conditions.

Heterotrophic Plate Counts

Heterotrophic plate counting was done using standard microbiological methods spread plate technique on R2A agar as described in *Standard Methods for the Examination of Water and Wastewater* (19th edition). Serial dilutions were prepared from each sample to yield 30 to 300 colonies per plate. A 100-µL sample from an appropriate dilution was withdrawn and used to inoculate the surface of the agar plate. A sterile bent glass rod was used to distribute the inoculum over the surface of the agar. Plates were incubated at room temperature for 7 days and then enumerated using a colony counter.

Acridine Orange Direct Counts

Acridine orange direct counts (AODC) enumerate all microbial cells in the samples. Analysis was completed at Indiana University Kokomo. AODC was performed by thoroughly vortexing the contents of one coupon with 10 mL of deionized water containing 0.01% (w/v) Tween 20 (on occasion the solution may have needed to be further homogenized with a mechanical homogenizer). Serial dilutions were prepared from each sample. A 1-mL sample from each appropriate dilution was withdrawn and mixed with 1-mL of a 0.1% (w/v) solution of acridine orange for five minutes at 20°C. The samples were then filtered on black cellulose nitrate filters, and washed once with 95% ethanol. The filters were observed by epifluorescence microscopy (Olympus BX-60). Fluorescent (orange to green) bacterial cells were counted in 25 different randomly chosen fields. The counting was automated using an image analysis system and a digital CCD camera.

Live/Dead Cell Enumeration

Analysis was completed at Indiana University Kokomo. Serial dilutions were prepared from each sample. A 1-mL sample from an appropriate dilution was withdrawn and mixed with 3 μL of a solution of SYTO 9/propidium iodide stains (Kit L-7007 Live/Dead® from Molecular Probes). This suspension was incubated in the dark at 20°C for 15 minutes. The samples were then filtered on a black cellulose nitrate filter and washed once with PBS. The filters were observed by epifluorescence microscopy (Olympus BX-60). Fluorescent red (dead) and green (live) bacterial cells were counted in at least 10 different randomly chosen fields. The counting was automated using an image analysis system and a digital CCD camera.

Chemical Samples

Water samples for chemical analysis were collected on a routine basis (every two weeks) at the effluent of the annular reactors as well as various locations prior to the inlet of the reactors (tap water, post BAC, pre UV, and post UV). Samples were collected in pre-cleaned 20 or 40-mL glass as appropriate.

Non-purgeable organic carbon (NPOC), trihalomethanes (THMs), haloacetic acids (HAA), organic acids, and aldehyde analysis were performed on collected samples.

Non-purgeable Organic Carbon (NPOC)

Samples were collected headspace-free in 40-mL pre-cleaned glass vials and preserved with concentrated phosphoric acid. Measurements were performed with a TOC-V CHP analyser (Shimadzu Corporation, Kyoto, Japan). Samples for DOC analysis were filtered through 0.45 μm pore-size membrane (Cole-Parmer® Nylon Membranes) rinsed 5 times with deionized water (APHA, AWWA, and WEF 1995).

Trihalomethanes

THM samples were collected headspace-free in 20-mL glass vials and preserved with ammonium chloride and acidified to pH 4.5 with hydrochloric acid. THM analysis was conducted as per EPA Method 551 at the University of Toronto. Gas chromatographic analyses of THMs were performed using a Hewlett Packard 5890 Series II-Plus GC, equipped with a DB-5 column for primary analysis, and a DB-1701 column for confirmation. A Fisons Mass Spectrophotometer (Trio 1000) was periodically used for compound identification.

Haloacetic Acids

HAA samples were collected headspace-free in 20-mL glass vials and preserved with ammonium chloride. HAA_6 was measured according to EPA Method 552.2 at the University of Toronto, which involves derivatization of the acids with acidified methanol. HAA_6 standards were obtained as a mix from Supelco®. Gas chromatographic analyses of HAAs were performed using a Hewlett Packard 5890 Series II-Plus GC, equipped with a DB-5 column for primary analysis, and a DB-1701 column for confirmation. A Fisons Mass Spectrophotometer (Trio 1000) was periodically used for compound identification.

Organic Acids and Aldehydes

During Phase 2 water samples were analyzed for carboxylic acids and aldehydes following UV/Cl_2 photocatalysis and in the effluent of the AR. An ion chromatographic method (Peldszus et al. 1996, Peldszus et al. 1998) was used to analyze the water samples for carboxylic acids at the University of Toronto. Without any preparation, samples were injected directly into the ion chromatograph using a large sample loop (740 μL). Utilizing a sodium hydroxide gradient (8 mM/10 min - 13 min - 125 mM/10 min) the organic acids were separated on an anion exchange column (AS 10, Dionex, Sunnyvale, CA) with conductivity detection. Low background conductivity was achieved by an anion self-generating suppresser (ASRS I, 4 mm, Dionex, Sunnyvale, CA), which was operated in the external water mode because of the relatively high sodium hydroxide concentration at the end of the gradient (125 mM).

Water samples for aldehyde analysis were derivatized overnight at room temperature with 1 mL σ-(2,3,4,5,6-pentafluorobenzyl)-hydroxylamine hydrochloride (PFBHA) and preserved with 32.5 mg of mercuric chloride ($HgCl_2$) and 32.5 mg of ammonium chloride (NH_4Cl). The aldehydes were then extracted from the sample with hexane as described by Sclimenti *et al.* (1990) and analyzed by gas chromatography equipped with electron capture detection (GC/ECD) (HP 5890 series II, Hewlett-Packard; Sunnyvale, CA).

Additional Analyses

Temperature and pH were measured on a regular basis using an Orion model 230A pH meter.

Disinfectants

Chlorine Dioxide. Chlorine dioxide was generated according to a method which was developed following a review of designs presented in *Standard Methods for the Examination of Water and Wastewater*, 19th ed. (APHA, AWWA, and WEF 1995), a method used by Sterling Pulp Chemicals, Ltd. (Toronto), and a method used by the Pulp and Paper Research Institute at the University of Toronto. The generator consists of a bench-top apparatus, in which a 25% sodium chlorite solution is slowly added to a 12N sulfuric acid solution, producing chlorine dioxide. The chlorine dioxide is purged from the mixture in a gas-washing bottle, and trapped in water surrounded by an ice bath. Off-gas from this bottle is directed to a potassium iodide trap to prevent the release of chlorine dioxide into the air. The resulting chlorine dioxide solution is approximately 2 g/L. The stock solution is standardized by mixing with a potassium iodide solution, and the titrating with sodium thiosulfate, first at pH 7.6 to measure one-fifth of the chlorine dioxide and any chlorine which is present, and then adding sulfuric acid to continue the titration at pH 2 to measure the remaining four-fifths of the chlorine dioxide plus any chlorite that was initially present. Past measurements indicate that the stock solution produced is essentially free (>99%) of chlorine species other than chlorine dioxide. Chlorine dioxide residuals were measured using a spectrophotometric method employing lissamine green as previously described by Chiswell and O'Halloran (1991).

Chlorine and Monochloramine. Concentrated disinfectant solutions of chlorine and monochloramine were produced using deionized water. Chlorine stock solution (about 100 mg/mL free chlorine) was prepared by adding 0.37 mL of sodium hypochlorite (6% solution, conforms to APHA and ASTM requirements) to 100 mL of PBS (pH 9.50). Ammonium chloride stock solution was prepared by adding 80 mg of NH_4Cl to 100 mL of PBS (pH 9.50). Monochloramine solution (about 100 mg/L as chlorine) was produced as needed by mixing equal volumes of each stock reagent. These disinfectants were diluted in PBS (pH 7.4) as required. Chlorine and monochloramine residuals were measured by the DPD ferrous titrimetric method (Standard Method, 19th edition - 4500 D).

Statistical Analysis

Results from the experiments were analyzed using analysis of variance (ANOVA) tests. ANOVA tests were used to examine the effect various experimental factors (e.g., chemical disinfectant dose) had on the number of biofilm and suspended HPCs. Unless otherwise noted, the level of significance (α) that were used for the statistical tests was 5%. For analysis purposes, all of the experimental factors were treated as qualitative variables.

CHAPTER 4
HYDRAULIC CALIBRATION AND FLUENCE DETERMINATION OF A MODEL UV DISINFECTION SYSTEM

INTRODUCTION

The use of potassium ferrioxalate to determine fluence has been shown to be effective and has been widely used because of its sensitivity over a wide spectral range and relative ease of use (Harris et al. 1987). The chemical was found to provide reproducible dose estimates but is sensitive to procedural variation. Chemical actinometry involves photochemical decomposition of known quantum yield when exposed to UV radiation at 253.7 nm and unlike bioassays provides relatively quick estimates of dose.

Havelaar et al. (1990) identifies the following reasons for choosing MS 2 for UV inactivation studies.

1. MS 2 is a single-strand RNA virus and consequently its dose-response curve of UV inactivation follows first order kinetics.
2. The structure and size of MS 2 is similar to that of human enteroviruses.
3. MS 2 has a relatively high UV resistance, comparable to that of bacterial spores.
4. MS 2 is not pathogenic to man, so it can be used for calibration of full-scale reactors without additional safety measures.

Several factors that influence the accurate estimation of fluence have been identified by Harris et al. (1987): flow rate through the reactor; short-circuiting of water as it traverses the complex geometries of many reactors; reductions in UV transmittance through reactor tube walls caused by scale build-up on the walls; UV absorbance of the water; age of the germicidal lamps; and ambient temperature around the germicidal lamps. The focus of the present investigation was the reactor hydraulics, which has been found to be of significant importance to the determination of fluence (Lyn and Blatchley 1999).

The primary objective of this study was to determine the effect that flow dynamics have on the performance of potassium ferrioxalate as a chemical actinometer. Ultraviolet fluence was evaluated at several flow rates and calibrated using a bioassay with MS 2 bacteriophages. MS 2 was used to obtain a dose response curve for the reactor. A goal of at least 2-log inactivation of MS 2 was chosen since MS 2 is known to be a UV-resistant organism. The final objective was to quantify the dispersion occurring at flow rates investigated for fluence determination.

CALIBRATION OF MODEL UV SYSTEM

The model UV system has previously been described in detail in Chapter 2. Additional information is supplied now to clarify the methods used in this study. The spectrum of the ultraviolet light emitted from the medium pressure lamp ranges from 200 to 450 nm with predominant emissions between 200 and 300 nm and emission spikes at 254, 312, 365, 405, and 435 nm as shown in Figure 4.1.

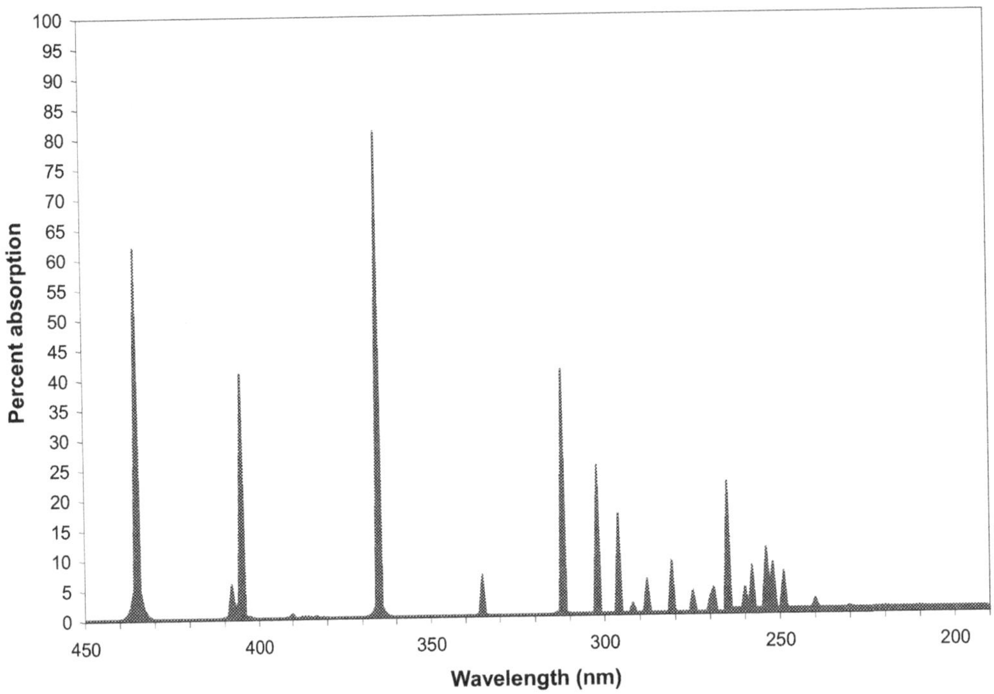

Figure 4.1 Absorbance spectrum for medium pressure UV lamp

Distilled water was pumped through the reactor with a variable speed modular peristaltic pump and master flex precision tubing (Saint-Gobain Performance Plastics, Wayne, NJ). The flow rates examined for this investigation were in the range of 100 – 300 mL/min. All joints and fittings used were either kynar or brass to minimize reactions with the potassium ferrioxalate solution and background ferrous iron concentrations. A static in-line pipe mixer containing 6 mixing elements (Koflo Corporation; Cary, IL) was installed just prior to the inlet of the quartz tube to assist with mixing before application to the quartz tube. The mixer is constructed from stainless steel and is 2.54 cm in diameter and 22.9 cm in length.

Calibration of the UV disinfection system was completed as summarized in the flow chart illustrated in Figure 4.2. The National Water Research Institute (NWRI) released a report on ultraviolet disinfection (Blatchley et al. 2000), which states that when using MS 2 as a bioassay "the data shall be conducted by plotting the dose-response data on a graph of the UV dose (mJ/cm^2) (x-axis) versus the log inactivation (y-axis)." The data must fall in the area bounded by Equations 4.1a and 4.1b when using a collimated beam apparatus.

$$-\log_{10}(N/N_o) = 0.040[\text{fluence, mJ/cm}^2] + 0.64 \qquad (4.1a)$$

$$-\log_{10}(N/N_o) = 0.033[\text{fluence, mJ/cm}^2] + 0.20 \qquad (4.1b)$$

The flow chart in shown in Figure 4.2 summarizes the calibration procedure that was used to obtain a fluence-response model for the described experimental system. As shown in Figure 4.2, a dispersion model was used to correct for non-ideal flow through the quartz tube. As

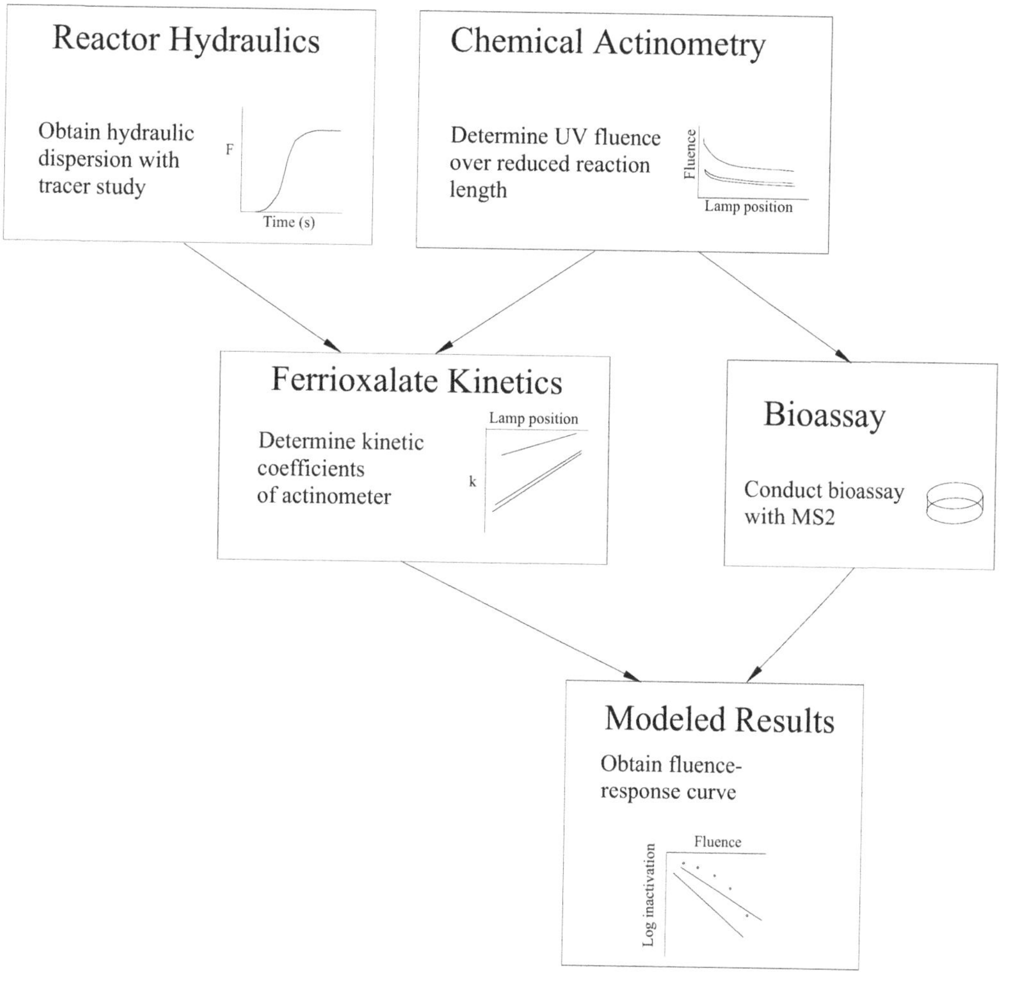

Figure 4.2 Calibration flow chart for experimental UV system (*Source*: Dykstra et al. [2002]. Reprinted from Hydraulic Calibration and Fluence Determination of Model UV Disinfection System, *ASCE – J. Enviro. Engrg.*, by permission. Copyright © 2002, ASCE.)

well an initial test was conducted using a chemical actinometer to determine whether the quartz tube should be covered and the height at which to operate the UV lamp. After these initial calibration procedures were conducted the reaction kinetics for the chemical actinometer was determined and the MS 2 inactivation experiments were performed. Subsequently, a fluence-response model could be obtained, as shown in Figure 4.2. The details of each mathematical and experimental method used in the calibration procedure are described herein.

Reactor Hydraulics

The hydraulic characteristics of the reactor were evaluated using a step tracer technique (Levenspiel 1999). Hydraulic retention time was determined by finite difference using Equation 4.2:

$$t_m = \sum_1^n t \frac{d\left(\frac{v}{m} \bullet C\right)}{dt} \Delta t \qquad (4.2)$$

where ν = fluid flowing through a vessel of volume V (m^3/s)
m = tracer (kg/s)
C = concentration of tracer (M)
t = time (s)

Chloride was used as a tracer for all experiments. Potassium chloride was prepared at a concentration of 3 mM because it represents a maximum conductance for the electrode. Hydraulic experiments were conducted at room temperature and at flow rates of 100, 130, 210 and 300 mL/min. Distilled water was allowed to flow through the reactor. At time $t = 0$, the KCl solution was applied. The Cl^- tracer concentration in the effluent was monitored using an electrode and conductivity meter (YSI, Yellow Springs, OH). Tracer concentration of 0.003 M was chosen because it represents a maximum conductance for the electrode used.

Transformation of the experimental C_{step} curve to an F curve was accomplished with Equation 4.3 (Levenspiel 1999):

$$F = \frac{v}{m} \bullet C_{step} \qquad (4.3)$$

where C_{step} = concentration of tracer (M)

Transformation of the F curve to the normalized E curve is accomplished using Equation 4.4.

$$E = \frac{dF}{dt} \qquad (4.4)$$

The Peclet number (Pe) can then be calculated from the E curve. In particular, the following steps were adapted from Weber and DiGiano (1995) to determine the Peclet number.

1. Determine t_m (mean value of the residence time) using Equations 4.5a and 4.5b.

$$t_m = \int_0^\infty tE(t)dt \qquad (4.5a)$$

$$t_m = \sum_1^n tE_t \Delta t \qquad (4.5b)$$

2. Determine Peclet number (Pe) using Equation 4.6.

$$\sigma^2_{\theta^o} = \frac{2}{Pe} - 2\left(\frac{1}{Pe}\right)^2\left[1 - \exp\left(-\frac{1}{Pe}\right)^{-1}\right] \qquad (4.6)$$

where $\sigma^2_{\theta^o} = \dfrac{\sum_{1}^{n} t^2 E_t \Delta t - t_m^2}{t_m^2}$, and

$$Pe = \frac{u\,L}{D}$$

Fluence Determination with Chemical Actinometry

Five positions of the UV lamp were investigated to determine the increase in dose for the various flow rates. Positions included both vertical extremes and three addition positions equidistant from the extremes.

Bulk quantities of 6 mM potassium ferrioxalate ($K_3Fe(C_2O_4)_3 \cdot 3H_2O$) were prepared in 4 L amber bottles using distilled water as described by Jagger (1967). The bottle was capped with a rubber stopper before application to the UV lamp. Solutions were prepared in amber bottles in order to limit exposure to light before UV irradiation. The solution was pumped to the UV lamp at flow rates of 100, 125, 130, 150 and 300 mL/min. Various flow rates were used to investigate retention time and hydraulic regime on actinometer photoreduction under conditions of different UV fluence. Influent and effluent samples were collected from inline sampling ports in a darkened syringe to limit light exposure and transferred to amber sample bottles for analysis. Sample preparation and measurement were conducted as reported by Jagger (1967).

UV fluence was calculated with Equation 4.7, as described by Jagger (1967) and reported by Harris et al. (1987).

$$Fluence(mJ/cm^2) = \frac{[Fe^{2+}]_a - [Fe^{2+}]_b}{\Phi} \times \frac{4.719 \times 10^8\, mW - s}{einstein} \times \frac{L}{10^3\, cm^3} \qquad (4.7)$$

where Φ = 1.26 moles Fe^{2+}/einstein

$[Fe^{2+}]_a$ = sample concentration of Fe^{2+} after irradiation (mol/L)

$[Fe^{2+}]_b$ = sample concentration of Fe^{2+} before irradiation (mol/L)

Equation 4.7 assumes that 100 percent of the radiation absorbed by the actinometer is at 253.7nm and that the quantum yield at this wavelength is 1.26 moles Fe^{2+}/einstein (Harris et al.

1987, Jagger 1967). This is not achieved in practice with a medium pressure UV lamp as the spectrum emitted ranges from 200 to 450 nm with predominant emissions between 200 and 300 nm.

Bioassay with MS 2

MS 2 bacteriophage (ATCC 15597-B1) propagation and enumeration were conducted using *Escherichia coli* C3000 (ATCC 15597) as a host. Amplification of MS 2 phages was done by infecting early-log phase host cells at 37°C for 16-24 hours. Phages were purified by centrifugation (10,000 x g, 10 minutes) and filtration through a 0.45-µm filter to remove bacterial host cells and debris. A plaque assay was utilized for enumeration. The assay was modified from the method described by the ISO (ISO 1995). The host cells were cultured for 3 to 4 hours in a tryptone broth (in distilled deionised water, 1.0% (w/v) tryptone, 0.1% (w/v) yeast extract, 0.1% (w/v) glucose, 0.8% (w/v) NaCl, and 0.0022% (w/v) $CaCl_2$). *E. coli* cell suspensions and MS 2 phage samples (undiluted or diluted in a calcium saline solution) were mixed together in molten tryptone agar (tryptone broth with 0.7 % (w/v) agar), and overlaid onto a bottom agar plate (tryptone broth with 1.2% (w/v) agar). Tryptone, yeast extract, and agar were purchased from Difco Laboratories (Detroit, MI, USA). Plates were incubated at 37ºC for 12 hours at which time the number of plaques was counted. Counts were replicated for each samples and variability between plaque counts for the same sample were approximately less than 10%.

MS 2 at a concentration of 1.0×10^5 pfu/mL were fed into the UV system. The phages were pumped into the reactor continuously at three flow rates (100, 125, and 150 mL/min) and allowed to reach steady-state as determined from tracer studies completed previously. The quartz tube was covered leaving a gap of 0.2, 0.4, and 0.6 cm for the three experiments completed.

RESULTS AND DISCUSSION

Hydraulic Calibration

Plug flow is assumed to exist in a closed tube reactor. Deviations from plug flow results from short-circuiting by recycle or stagnant zones in the reactor (Weber and DiGiano 1995). It is, therefore, necessary to determine reactor flow and mixing characteristics experimentally. The C_{step} curve that was generated from the tracer study was converted to an F curve and normalized to an E curve. The E curve illustrates the residence time distribution (RTD) in the reactor. Transformation of the experimental C_{step} curve to an F curve was accomplished by the procedure described by Levenspiel (1999). Transformation and dispersivity calculations for the UV reactor are provided in Equations 4.3–4.7.

The data were analysed to calculate the theoretical hydraulic retention time (HRT) of the reactor (volume/flow rate); the mean residence time (t_m); the variance (σ_t); the skewness (s); the dimensionless form of variance (σ_θ); and Pe.

Figure 4.3 illustrates the results from the step tracer experiment at 130 mL/min and 210 mL/min without the mixer and with the mixer present. The figure illustrates the transformation from the experimental C_{step} to the F curve. The F curve represents the dimensionless form of the

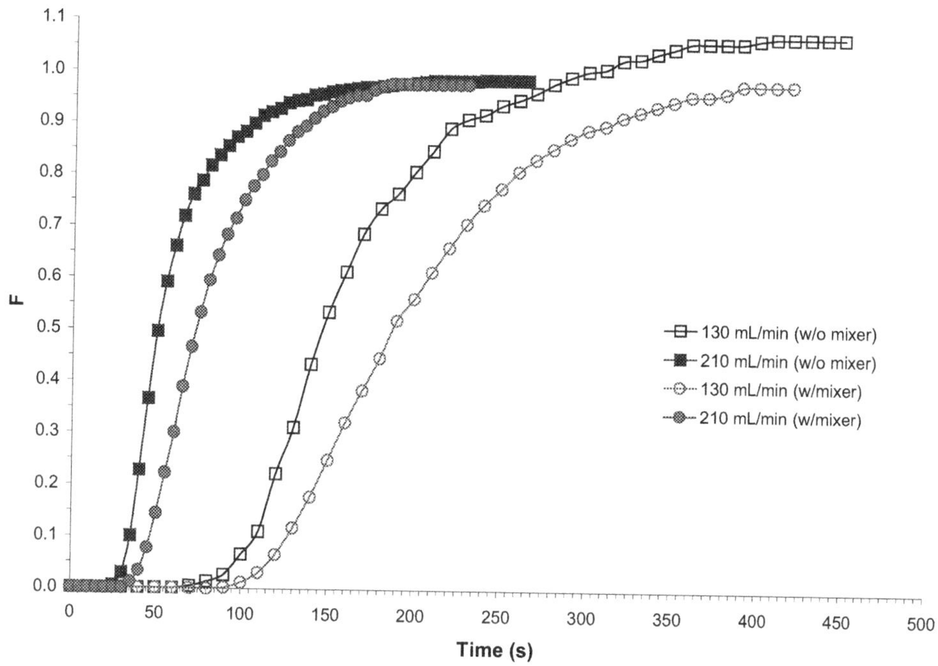

Figure 4.3 Experimental determined F curves for the system with and without the static mixer installed (*Source*: Dykstra et al. [2002]. Reprinted from Hydraulic Calibration and Fluence Determination of Model UV Disinfection System, *ASCE – J. Enviro. Engrg.*, by permission. Copyright © 2002, ASCE.)

C_{step} curve and is found by having the tracer concentration rise from zero to unity (Levenspiel 1999).

During tracer experimentation without the mixer it was noted that the tracer solution was being measured some time before the HRT of the quartz tube had passed. Further investigation with a visible tracer revealed the formation of a jet within the quartz tube. It was concluded that the drastic change in diameter from the influent supply tubing to the quartz tube was creating the jet. As a result, an in-line static pipe mixer was installed just prior to the influent of the reactor and the tubing inlet configuration was modified to minimize the jetting effect. Another tracer study was performed to determine if jetting in the reactor was still a problem.

Figure 4.3 indicates that there was little change in dispersion within the reactor. The jetting observed in the previous tracer study was eliminated as measured tracer solution occurred at the HRT of the reactor. This was confirmed with a visible tracer. An additional tracer experiment was conducted at a flow rate of 300 mL/min to enable dispersion quantification for chemical actinometry experiments completed previously. The Peclet number and dispersivity data for each flow rate and entrance condition are presented in Table 4.1.

Under all flow conditions, the skewness was greater than five (s_3>5). A positive value of s_3 indicates that the residence time distribution (RTD) is skewed towards residence greater than the mean value (Weber and DiGiano 1995). This parameter decreases with an increase in the volumetric flow rate. The Peclet number decreases with an increase in flow rate. This indicates that a larger amount of dispersive mixing is occurring at a lower flow rates because large values of Pe indicate that the relative influence of dispersion is increased (Bailey and Ollis 1986).

Table 4.1

Dispersion results from tracer study without in-line pipe mixer*

Flow (mL/min)	With Mixer t_m (s)	Pe	D (cm^2/s)	Without Mixer t_m (s)	Pe	D (cm^2/s)
100	268	22.7	1.3	-	-	-
130	190	13.5	2.88	176	24.4	1.58
210	76	8.9	7.05	54	3.3	18.8
300	59	8.6	10.3	-	-	-

Table 4.1 illustrates that the dispersivity is different for each flow regime for tests conducted with and without the mixer. It is hypothesized that the jet formation observed when the mixer was absent had a more pronounced effect on the dispersion than that observed for the lower flow rate. As a result the Peclet number increased at the lower flow rates. Under all flow conditions the Peclet number was greater than 8.6 except where short-circuiting was a factor at the higher flow rate.

Analysis of Chemical Actinometry

Preliminary investigation of the fluence supplied to the quartz tube resulted in fluences that were well beyond the range that is required for further studies in water treatment. Higher flow rates were investigated to reduce the fluence to the desired range of 30 to 80 mJ/cm^2. Increasing the flow was effective at lowering the fluence but did not produce expected results. Ultraviolet fluence increased linearly with changes in lamp position at a constant flow rate. However, changes in flow rates did not indicate a trend at constant lamp position. Figure 4.4 illustrates a noticeable difference in the linear relationship with a corresponding change in flow rate.

At the higher flow rate the corresponding change in fluence with respect to lamp position did not change to the same extent as the lower flow rate. The tracer study revealed the formation of a jet in the reactor, which was increasingly apparent as the flow rate increased. The relationship between the fluence and flow rate could have been affected by the formation of the jet.

This idea is supported with results from experimentation completed after the in-line pipe mixer was installed. Figure 4.5 illustrates that at the higher flow rate the corresponding change in fluences with respect to lamp position is similar to change observed at the lower flow rate. This result was expected during the first set of experiments completed before the in-line pipe mixer was installed. The UV fluence applied in Figure 4.5 ranged from 200 to 3000 mJ/cm^2, which is outside the practical range for water treatment. Therefore, further experiments were needed to optimize the UV fluence to values more typical of water treatment.

Comparison between Figures 4.4 and 4.5 also shows that UV fluence determination with potassium ferrioxalate appeared to be affected by the hydraulics in the reactor. The presence of the jet appears to play an important role in quantifying the fluence with potassium ferrioxalate. In contrast Harris et al. (1987) conclude that reactor hydraulics were not expected to affect the performance of potassium ferrioxalate as a chemical actinometer because the photoreduction

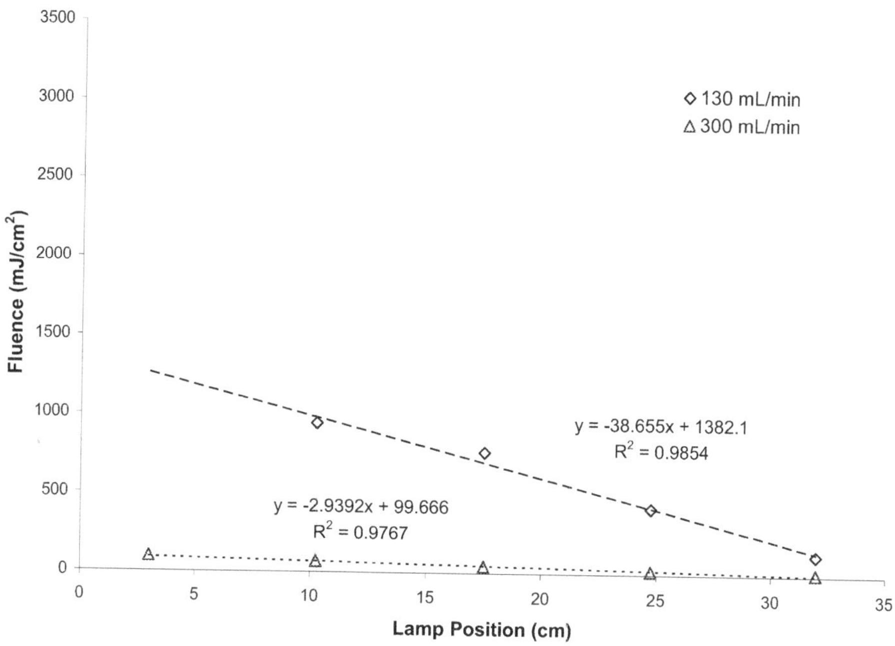

Figure 4.4 **Linear relationship between lamp position and fluence at constant flow rates before installation of an in-line mixer (over the entire length of the quartz tube) (*Source*: Dykstra et al. [2002]. Reprinted from Hydraulic Calibration and Fluence Determination of Model UV Disinfection System, *ASCE – J. Enviro. Engrg.*, by permission. Copyright © 2002, ASCE.)**

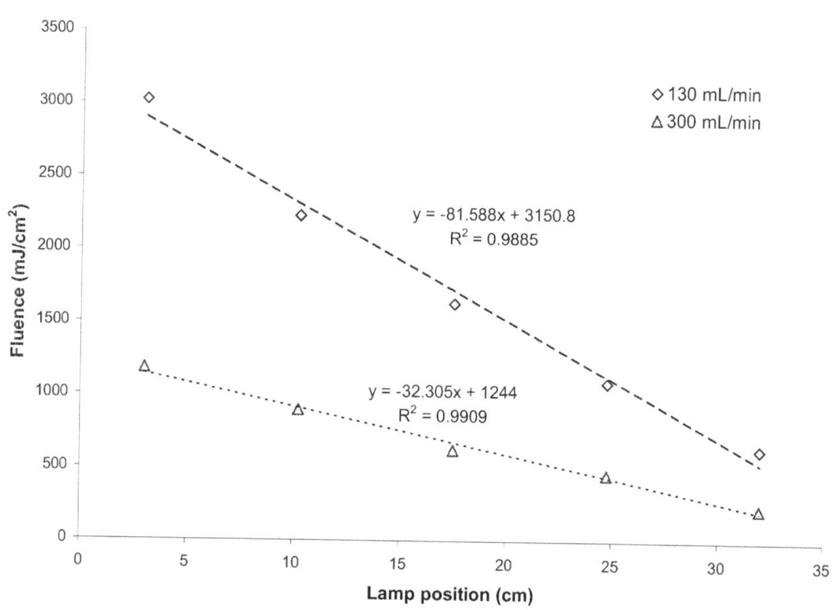

Figure 4.5 **Linear relationship between lamp position and fluence at constant flow rates after the installation of an in-line mixer (over the entire length of the quartz tube) (*Source*: Dykstra et al. [2002]. Reprinted from Hydraulic Calibration and Fluence Determination of Model UV Disinfection System, *ASCE – J. Enviro. Engrg.*, by permission. Copyright © 2002, ASCE.)**

of the actinometer is zero order. Consequently, batch data from collimated beam tests can be applied directly to continuous flow reactors regardless of their flow regime. This was not found to be the case in this study as reactor hydraulics appears to affect the performance of the actinometer.

To decrease the fluence to between 20 and 150 mJ/cm^2, the quartz tube was covered on either end with aluminium foil to a spacing of 0.5 cm at the centre. Experiments performed with potassium ferrioxalate resulted in fluences between 50 and 225 mJ/cm^2, as shown in Figure 4.6. With the decrease in exposure time there appears to be a correlation between the flow rates that was not previously observed. It appears that the actinometer is more sensitive to changing lamp position. As the lamp approaches the quartz tube the fluence will approach infinity. This was not observed during experimentation over the entire length of the tube. With this arrangement the fluence predicted by chemical actinometry may be overestimated.

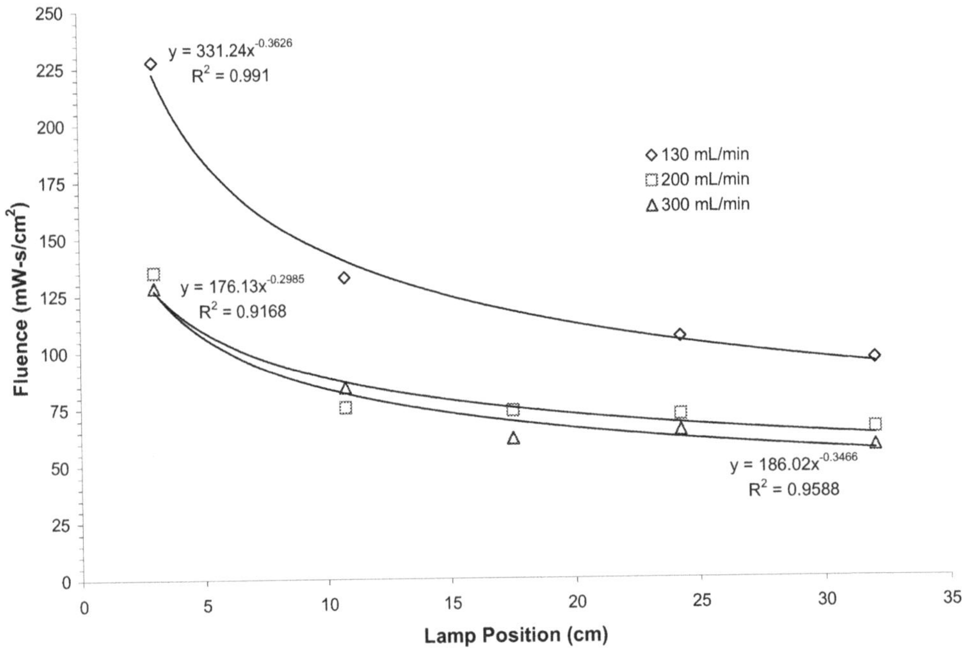

Figure 4.6 Fluence at three flow rates with a 0.5-cm space over the length of the quartz tube (*Source*: Dykstra et al. [2002]. Reprinted from Hydraulic Calibration and Fluence Determination of Model UV Disinfection System, *ASCE – J. Enviro. Engrg.*, by permission. Copyright © 2002, ASCE.)

Kinetic Rate Determination and Model Calibration

To account for the dispersion in the quartz tube, an advection-dispersion model, with zero order reaction, was used to determine the decay coefficient for the chemical actinometer for the two hydraulic regimes. The photolytic reaction is zero order and can be solved numerically as indicated in Equations 4.8 and 4.9, respectively:

$$\frac{\partial C}{\partial t} = D\frac{\partial^2 C}{\partial x^2} - v_x\frac{\partial C}{\partial x} - k \quad (4.8)$$

$$C_{i,n+1} = C_{i-1,n}\left[\frac{D\Delta t}{\Delta x^2} + \frac{v_x\Delta t}{\Delta x}\right] + C_{i,n}\left[1 - 2\frac{D\Delta t}{\Delta x^2} - \frac{v_x\Delta t}{\Delta x}\right] + C_{i+1,n}\frac{D\Delta t}{\Delta x^2} - k\Delta t \quad (4.9)$$

Figure 4.7 illustrates the trend in the negatively increasing decay coefficient as the lamp is positioned closer to the quartz tube. This is expected, as the actinometer would be photodegraded more quickly at a higher intensity (i.e. when the lamp is positioned closer to the quartz tube). The graph clearly shows that the decay coefficients at 130 mL/min (with and without the in-line mixer) are very different. The jetting effect observed without the mixer resulted in decay coefficients that are much lower than those observed with the mixer. There also does not appear to be an appreciable difference in the decay coefficients at the low and high flow rate with the mixer present.

Due to the dispersion that occurred within the quartz tube, an advection-dispersion model at steady state was used to compare it with design equations (4.1a and 4.1b) developed for a collimated beam configuration (Blatchley et al. 2000).

$$0 = D\frac{d^2 C}{dx^2} - v_x\frac{dC}{dx} - k \quad (4.10)$$

The analytical solution for Equation 4.10 can be obtained and is expressed as:

$$C = \phi\frac{D}{v_x^2}e^{\frac{v_x}{D}x} - \frac{k}{v_x}x + \alpha \quad (4.11)$$

where C = concentration of iron in solution (M)
D = dispersion (cm^2/s)
v_x = velocity of flowing fluid (cm/s)
k = decay constant (M/s)
x = photolytic reaction length (cm)
ϕ, α = numerical constants

The model presented in Equation 4.11 was evaluated under flow condition and chemical actinometry analysis over the entire length of the quartz tube. The reaction kinetics of potassium ferrioxalate was assumed to be constant for each lamp position, which was confirmed as presented in Figure 4.7. The model can be used to determine the concentration of actinometer that would exist for photolytic reaction lengths that are less than or equal to 42.2 cm (i.e. full reaction length). The resulting dose response curves for photolytic reaction lengths of 0.2, 0.4, and 0.6 cm are illustrated in Figure 4.8. The modeled results presented in Figure 4.8b for a photolytic reaction length of 0.4 cm was selected as the best calibration reaction length as the slope of the line is within the range of the NWRI design curves. The shorter reaction length illustrated in Figure 4.8a resulted in a slope that was greater than the range of the NWRI design curves. In contrast the longest photolytic reaction length produced a curve that had a slope approximately equal to the upper NWRI design curve (Figure 8c). Furthermore, at each reaction length the y-intercept was a positive value, which indicates that there is a minimum fluence value (UV_{min}) required to obtain log inactivation. At a photolytic reaction length of 0.4 cm, for example, the UV_{min} was calculated as 9.4 mJ/cm^2. While a positive y-intercept is not consistent with the NWRI curves, it provides a minimum value for log inactivation. The negative y-intercept for the NWRI curves implies that a UV fluence of 0 mJ/cm^2 would yield a 0.5-log inactivation. While it is recognized that the design curves are linear models generated from a large pool of data, it is physically not possible to obtain an inactivation of MS 2 without any UV fluence applied. Therefore it is recommended that care be used when calibrating a UV system to the NWRI design curves, particularly in the lower UV fluence range.

There are several possible explanations for the differences between the NWRI design curves and the modeled results from the present investigation. First, the design curves identified by Blatchley et al. (2000) were developed for a collimated beam apparatus not a flow through system used in this experiment. A flow through system introduces experimental errors resulting from the tracer study, fluence determination with chemical actinometry and bioassay experimentation, which would not be experienced to the same extent in a collimated beam test.

Another significant difference was that the present investigation used a medium pressure lamp that has a UV spectral range of 200 to 450 nm with emission spikes 254, 312, 365, 405, and 435 nm; whereas the NWRI curves were developed for a low pressure lamp with a monochromatic wavelength of 254 nm.

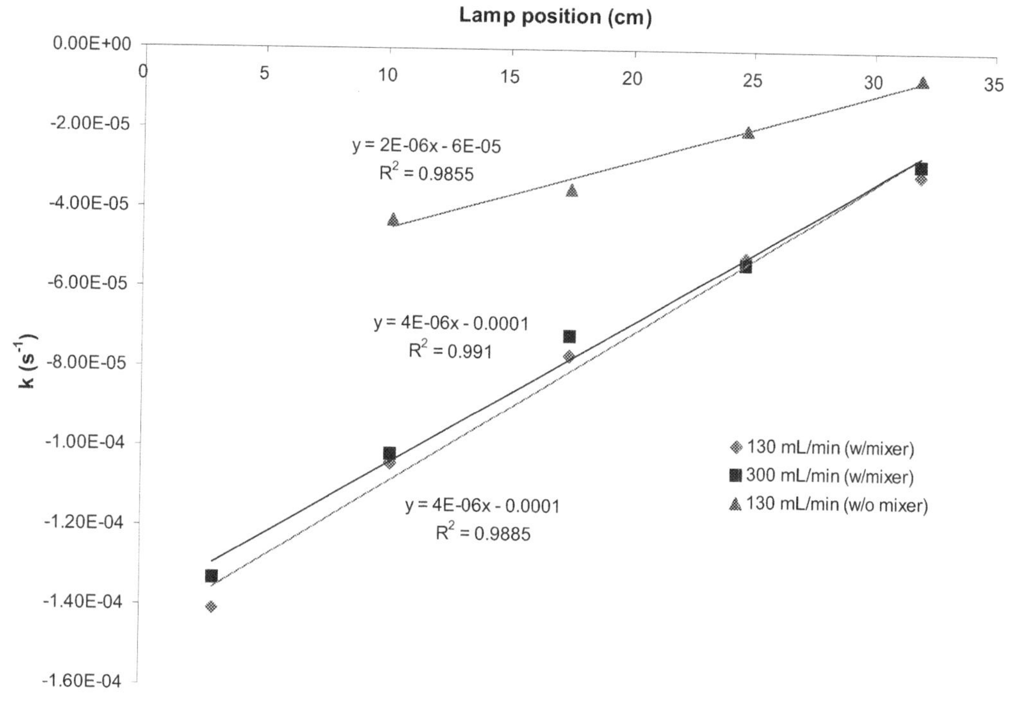

Figure 4.7 Ferrioxalate decay coefficients for various lamp positions (*Source*: Dykstra et al. [2002]. Reprinted from Hydraulic Calibration and Fluence Determination of Model UV Disinfection System, *ASCE – J. Enviro. Engrg.*, by permission. Copyright © 2002, ASCE.)

SUMMARY

This study investigated the impact hydraulic mixing has on the calibration of a flow-though model UV-disinfection system with chemical actinometry (potassium ferrioxalate). Under the experimental conditions studied, dispersion had a significant role in the ability for potassium ferrioxalate to accurately measure the fluence supplied to water flowing though a reactor. Under experimental conditions chemical actinometry consistently overestimated the fluence being supplied within the reactor because of poor mixing and short-circuiting within the reactor. To correct for these mixing problems, an axial-dispersion model was used to determine the rate kinetics of potassium ferrioxalate. After dispersion and kinetics are accounted for within the UV system, the modeled results for a photolytic reaction length of 0.4 cm predicted UV fluence that was in general agreement with UV design curves for inactivation MS 2 bacteriophage. The differences in the NWRI design curves and the fluence-response model in the present investigation were found to be related to the experimental errors introduced from using a flowing system and because a medium pressure lamp was used in the present investigation.

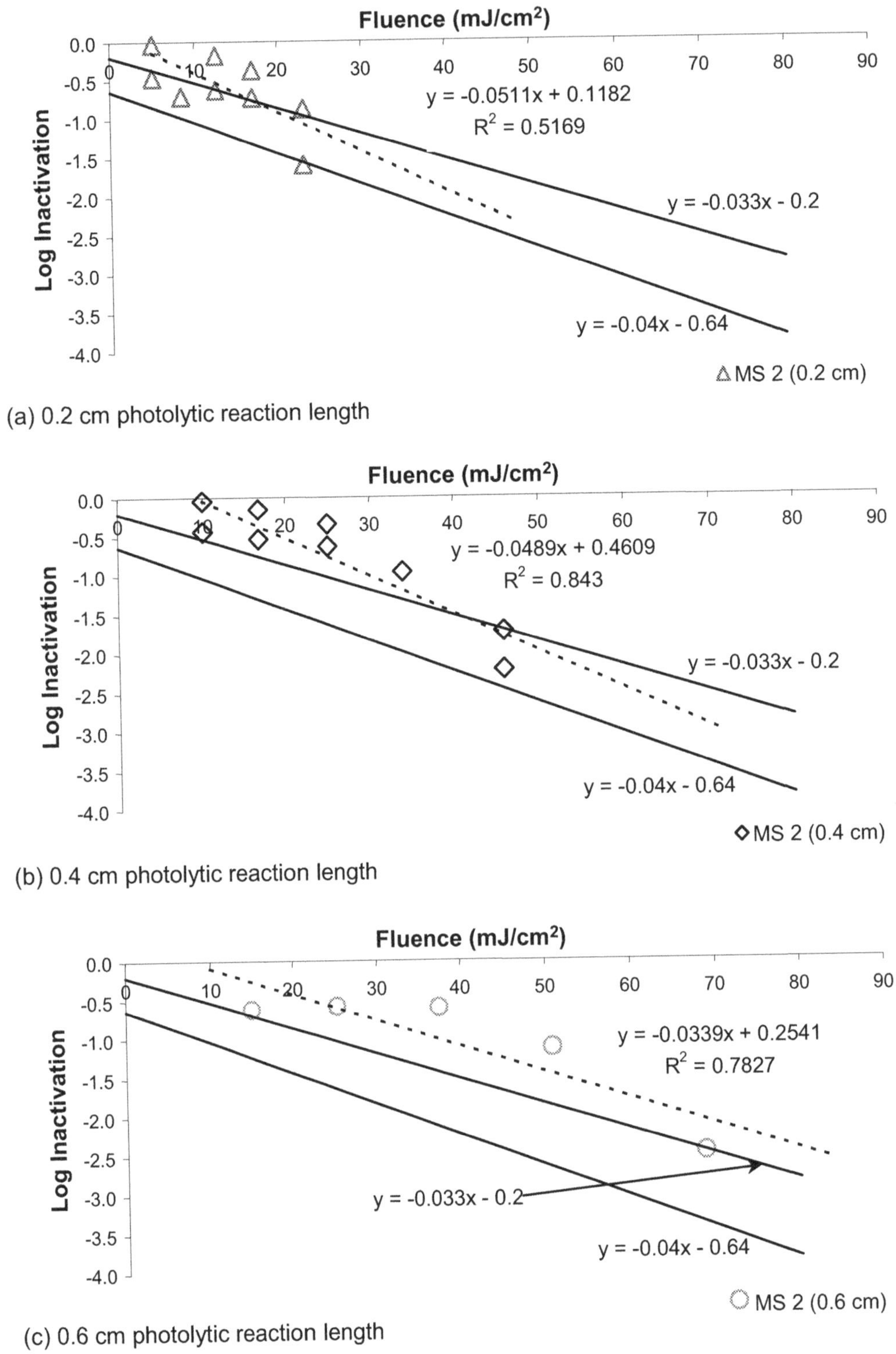

Figure 4.8 Modeled MS 2 dose response data with UV design curves shown for reference purposes (*Source*: Dykstra et al. [2002]. Reprinted from Hydraulic Calibration and Fluence Determination of Model UV Disinfection System, *ASCE – J. Enviro. Engrg.*, by permission. Copyright © 2002, ASCE.)

CHAPTER 5
IMPACT OF UV DISINFECTION AND DISINFECTANT CONCENTRATION ON BIOFILM ACCUMULATION (PHASE 1)

INTRODUCTION

The overall objective of this chapter was to evaluate the impact of UV disinfection on biofilm accumulation in drinking water distribution systems. In addition, the research compared the effectiveness of chlorine dioxide, free chlorine, and monochloramine as secondary disinfectants for reducing bacterial regrowth (bulk water and biofilm cells) in the presence and absence of UV disinfection. Previous studies have evaluated free chlorine and monochloramine for controlling biofilm growth, however there is a general lack of published data concerning chlorine dioxide in this area. Furthermore there is very little information concerning the potential synergistic effects that may control regrowth when applying UV at dosages required for *Cryptosporidium parvum* inactivation, prior to a chlorine-based disinfectant.

MATERIALS AND METHODS

Specific methods for the operation of the UV disinfection assembly and the annular reactors are described in detail in Chapter 3. Figure 5.1 illustrates the specific details related to the operation of the process water being passed through UV disinfection system and subsequently pumped into the reactors.

Disinfection commenced after the 4-week acclimation period was completed and a steady-state biofilm was achieved, as indicated by a relatively constant number of biofilm HPCs. Data collected during the subsequent disinfection period enabled a direct comparison of chlorine dioxide, free chlorine, and chloramines for controlling biofilm growth in distribution systems. One of the four ARs did not receive any disinfectant residual, as shown in the experimental matrix (Table 5.1). Experimental trials with a low and a high effluent disinfectant residual concentrations were performed on water that had been UV treated and water that had not been UV treated.

RESULTS AND DISCUSSION

Microbial Analysis

Heterotrophic Bacteria

For UV disinfection experimentation influent, the concentration of suspended heterotrophic bacterial cells (post-BAC column) for low and high chemical disinfection phases during the acclimation period had geometric means of 6600 and 1900 CFU/mL, respectively. Influent suspended bacterial cell concentrations (post-UV treatment) for low and high chemical disinfection phases during the acclimation period had geometric means of 3900 and 1500 CFU/mL, respectively. Finally, the concentration of bacteria during the acclimation period in the control AR for low and high chemical disinfection phases had geometric means of 3.66×10^4 and 5.18×10^4 CFU/mL, respectively.

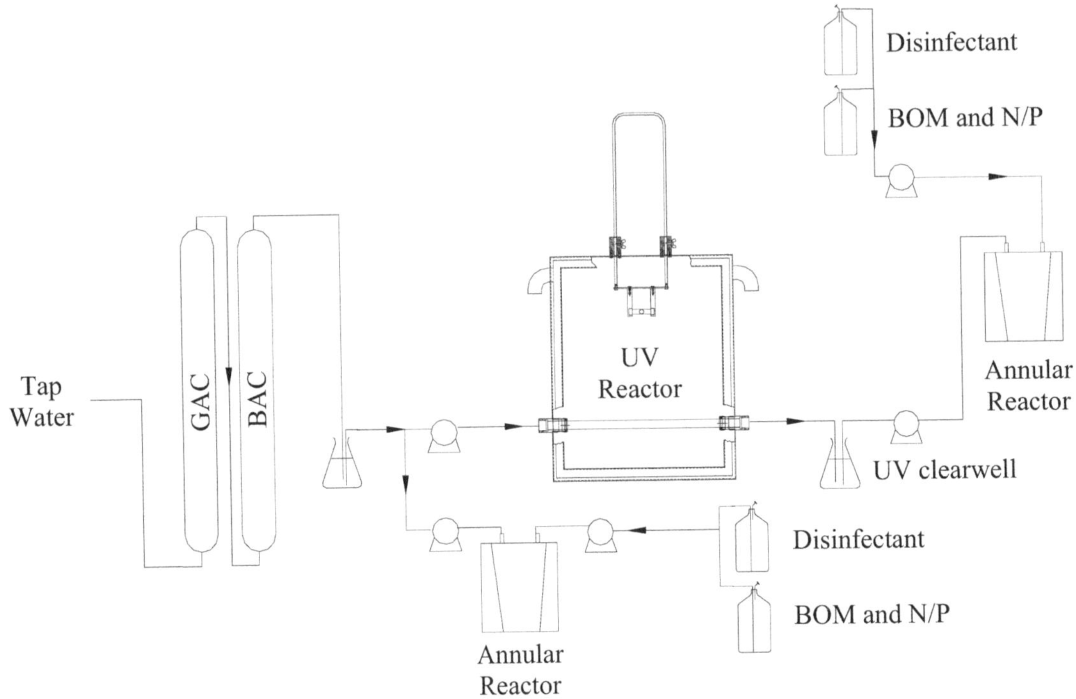

Figure 5.1 Schematic of bench-scale system for Phase 1

For experimentation without UV disinfection, influent suspended bacterial cell concentrations (post-BAC column) for low and high chemical disinfection phases during the acclimation period had geometric means of 6500 and 8700 CFU/mL, respectively. The bacterial concentration during the acclimation period in the control AR for low and high chemical disinfection phases had geometric means of 1.91×10^4 and 12.4×10^4 CFU/mL, respectively.

Suspended and biofilm heterotrophic plate counts (HPCs), shown in Figures 5.2 and 5.3 for the ARs that received UV disinfection and low disinfection residuals, show similar trends regardless of the treatment received. However, the ARs receiving chlorine dioxide and free chlorine had distinctly superior performance in comparison to the chloramines reactor. Disinfection in the ARs began after the fourth week of experimentation when a steady state biofilm was established, which is consistent with biofilm work of others (Camper 1996). Throughout experimentation the pH of the water has been 7.4 ± 0.4 and the temperature has been 20 ± 0.3 °C.

Table 5.1

Experimental matrix for Phase 1

Supplemental Disinfection	Chemical Disinfectant Residual Concentration (mg/L)			
	Control: No Disinfectant	Chlorine Dioxide	Free Chlorine	Monochloramine
With UV Disinfection	0.0/0.0	0.25/0.5[1]	0.5/1.0	1.0/2.0
Without UV Disinfection	0.0/0.0	0.25/0.5	0.5/1.0	1.0/2.0

[1] Low/High effluent chemical disinfection residual concentrations.

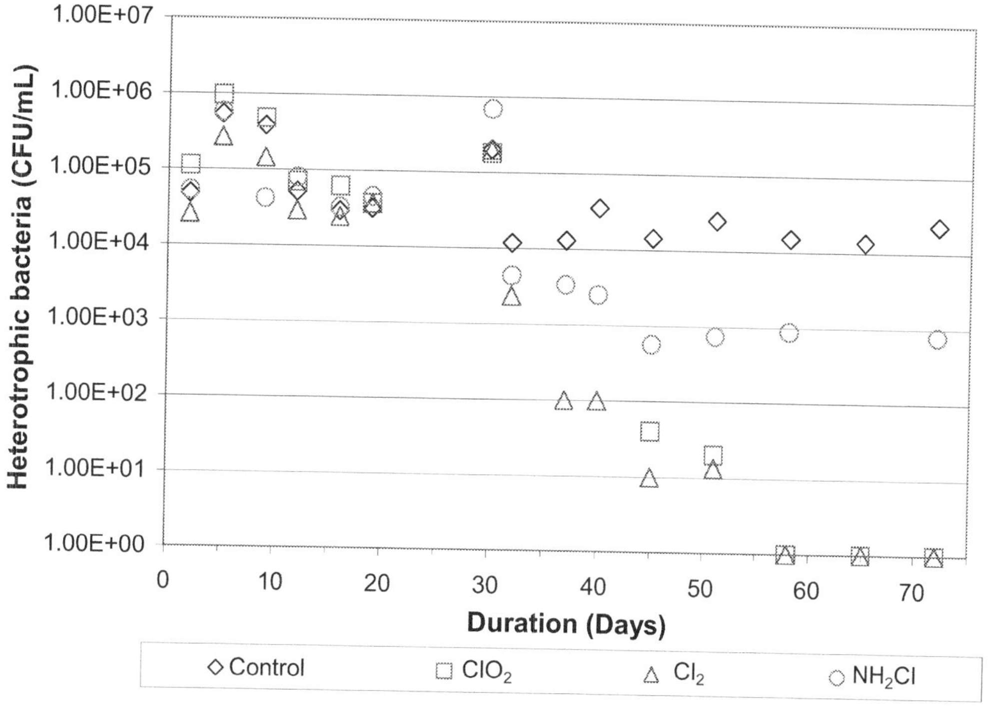

Figure 5.2 Suspended HPCs in ARs with UV disinfection (low chemical disinfectant residual concentrations)

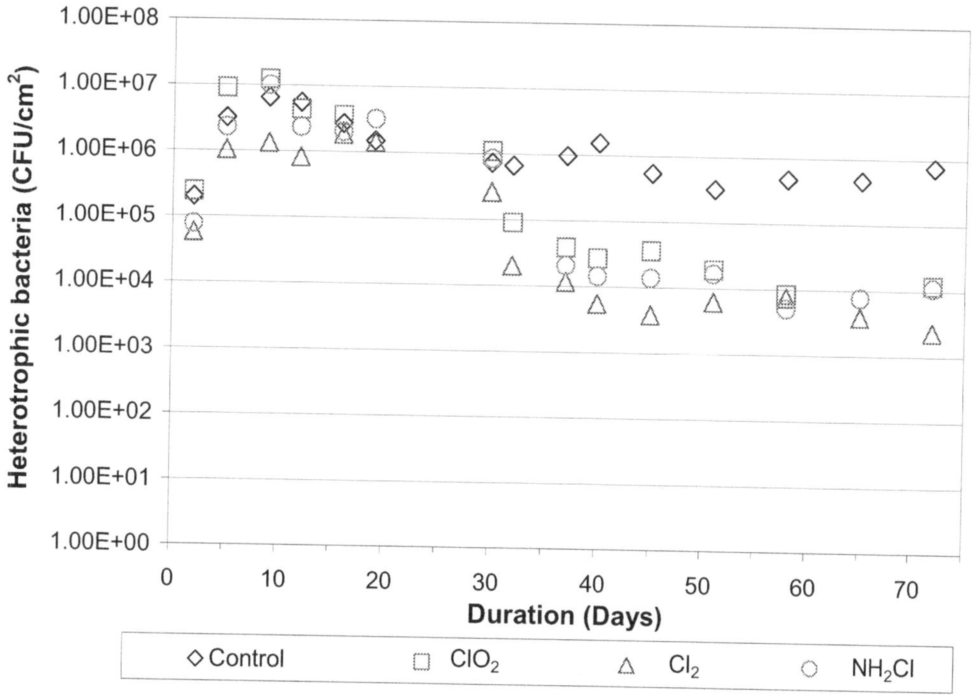

Figure 5.3. Biofilm HPCs in ARs with UV disinfection (low chemical disinfectant residual concentrations)

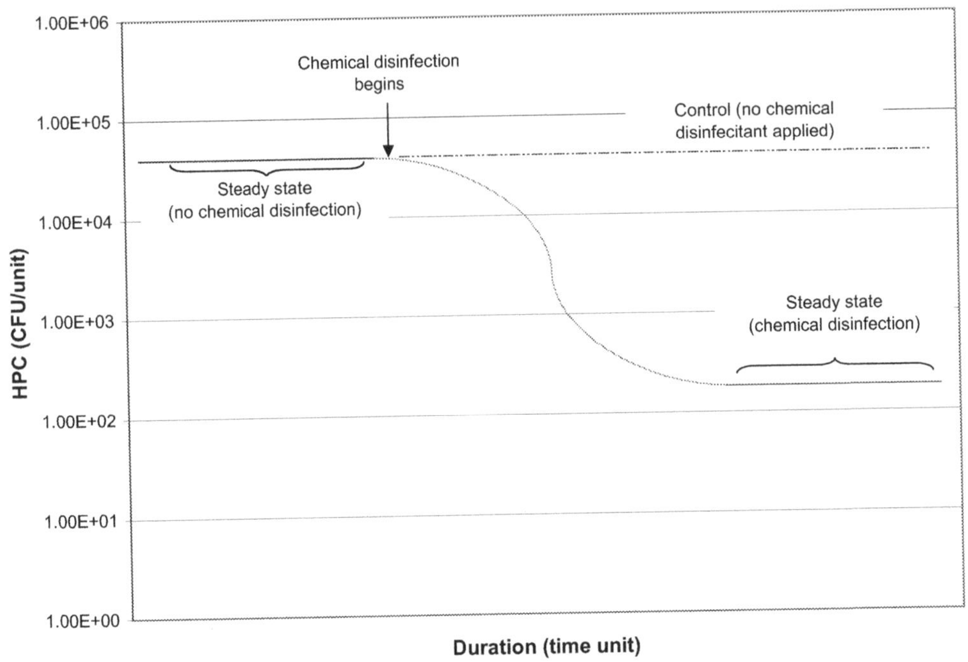

Figure 5.4 Hypothetical HPC counts over an experimental trial period

Figure 5.4 presents a hypothetical curve for HPCs over an experimental trial period. The figure illustrates that a pseudo-steady state period is achieved before the application of chemical disinfectants. The acclimation level of HPC is maintained throughout the experimental trial by the control AR because it does not receive any chemical disinfection. For the ARs that did receive chemical disinfection, an initial microbial inactivation period is quickly observed after the application of disinfectants and this period can last several days. A final pseudo-steady state period is eventually achieved for the disinfection period of the experimental trial. From this graphical approach log-reductions in HPC bacteria can be calculated by using Equation 5.1:

$$Log\ reduction = -\log\left(\frac{HPC_{disinf}}{HPC_{control}}\right) \tag{5.1}$$

Where HPC_{disinf} represents the pseudo-state number of HPC during chemical disinfection, and $HPC_{control}$ represents the pseudo-state number of HPC during no chemical disinfection.

Steady-state geometric mean values for suspended and biofilm HPCs prior to and following chemical disinfection are shown in Table 5.2 for UV disinfection experimentation and Table 5.3 for experimentation without UV disinfection. Geometric mean values for suspended and biofilm HPCs reflect the last four sample points just prior to disinfection and the last four sample points of the experimental trial when the bacterial concentrations had reached pseudo-steady state. Log reductions in heterotrophic bacteria for each of the disinfectants are also shown in Table 5.2 for UV disinfection experimentation and Table 5.3 for experimentation without UV disinfection.

Table 5.2

Geometric mean values for suspended and biofilm HPCs with UV disinfection

		No Chemical Disinfection HPCs		Chemical Disinfection HPCs		Log Reduction	
		Suspended	Biofilm	Suspended	Biofilm	Suspended	Biofilm
		× 10^4 CFU/mL	× 10^6 CFU/cm^2	× 10^4 CFU/mL	× 10^6 CFU/cm^2		
A	Control	3.66	2.00	1.87	0.48	0.29	0.62
	ClO_2	5.44	2.57	0.0006	0.013	3.95	1.59
	Cl_2	2.99	0.87	0.0004	0.005	3.99	2.02
	NH_2Cl	4.95	1.89	0.08	0.009	1.37	1.72
B	Control	5.18	1.84	4.94	1.06	0.02	0.24
	ClO_2	4.23	1.18	0.0001	0.009	4.69	2.08
	Cl_2	4.31	1.72	0.0001	0.005	4.69	2.32
	NH_2Cl	4.71	1.01	0.16	0.05	1.49	1.33

A – Low chemical disinfectant residual concentrations.
B – High chemical disinfectant residual concentrations.

Because the disinfectant concentrations are not the same for the "low" and "high" levels, comparisons of inactivation data were conducted on the basis of equivalent disinfectant concentration and contact time (Ct). Table 5.4a and Table 5.4b show the average disinfectant concentrations in the effluent AR and the calculated Ct for the ARs with UV treatment and without UV treatment.

Comparisons of disinfection efficacy were made by combining the log reduction data presented in Tables 5.2 and 5.3 with the Ct data presented in Table 5.4a and Table 5.4b. In Figure 5.5, for example, free chlorine and chlorine dioxide were compared at a Ct value of approximately 30 mg/L·min. Without UV treatment chlorine dioxide provided a reduction in suspended heterotrophic bacteria that was about 2-log greater than free chlorine for the same Ct. With UV treatment, the difference between free chlorine and chlorine dioxide appeared less for

Table 5.3

Geometric mean values for suspended and biofilm HPCs without UV disinfection

		No Chemical Disinfection HPCs		Chemical Disinfection HPCs		Log Reduction	
		Suspended	Biofilm	Suspended	Biofilm	Suspended	Biofilm
		× 10^4 CFU/mL	× 10^6 CFU/cm^2	× 10^4 CFU/mL	× 10^6 CFU/cm^2		
A	Control	1.89	1.04	2.52	0.27	-0.12	0.60
	ClO_2	2.21	2.78	0.062	0.13	1.61	0.30
	Cl_2	1.87	0.82	0.016	0.007	2.20	1.60
	NH_2Cl	2.13	1.34	0.093	0.036	1.43	0.86
B	Control	12.4	1.12	13.9	1.04	-0.05	0.03
	ClO_2	8.26	0.88	0.001	0.007	4.00	2.17
	Cl_2	8.90	0.75	0.008	0.004	3.25	2.44
	NH_2Cl	11.1	1.08	0.041	0.007	2.53	2.15

A – Low chemical disinfectant residual concentrations.
B – High chemical disinfectant residual concentrations.

Table 5.4a

Chemical disinfectant residual concentrations and Ct values with UV disinfection

		Target Residual Concentration (mg/L)	Average Measured Residual Concentration (Standard Deviation)	Ct[1] (mg/L·min)
A	Control	0.0	-	-
	ClO_2	0.25	0.21 (0.09)	13
	Cl_2	0.5	0.52 (0.13)	31
	NH_2Cl	1.0	0.81 (0.25)	48
B	Control	0.0	-	-
	ClO_2	0.5	0.43 (0.10)	26
	Cl_2	1.0	0.94 (0.14)	57
	NH_2Cl	2.0	1.53 (0.40)	92

suspended HPCs than it did for the ARs that did not have UV treatment. However, the HPC levels were below the detection limit with UV treatment and it is plausible that the log-inactivation levels for free chlorine and chlorine dioxide were greater than the value shown in Figure 5.5.

Similarly, free chlorine and chloramines were compared in Figure 5.6 for Ct levels of approximately 60 mg/L·min. At this inactivation level free chlorine resulted in a greater log-inactivation of biofilm and suspended HPCs, independent of UV treatment. However, UV treatment increased the difference between free chlorine and chloramine inactivation from 1.5-log to greater than 3.0-log for suspended HPCs.

Table 5.4b

Chemical disinfectant residual concentrations and Ct values without UV disinfection

		Target Residual Concentration (mg/L)	Average Measured Residual Concentration (Standard Deviation)	Ct* (mg/L·min)
A	Control	0.0	-	-
	ClO_2	0.25	0.23 (0.07)	14
	Cl_2	0.5	0.47 (0.06)	28
	NH_2Cl	1.0	0.97 (0.07)	58
B	Control	0.0	-	-
	ClO_2	0.5	0.45 (0.09)	27
	Cl_2	1.0	0.95 (0.08)	57
	NH_2Cl	2.0	1.85 (0.09)	111

A – Low chemical disinfectant residual concentrations.
B – High chemical disinfectant residual concentrations.
* – Ct calculated from the average measured chemical residual concentration (time in minutes).

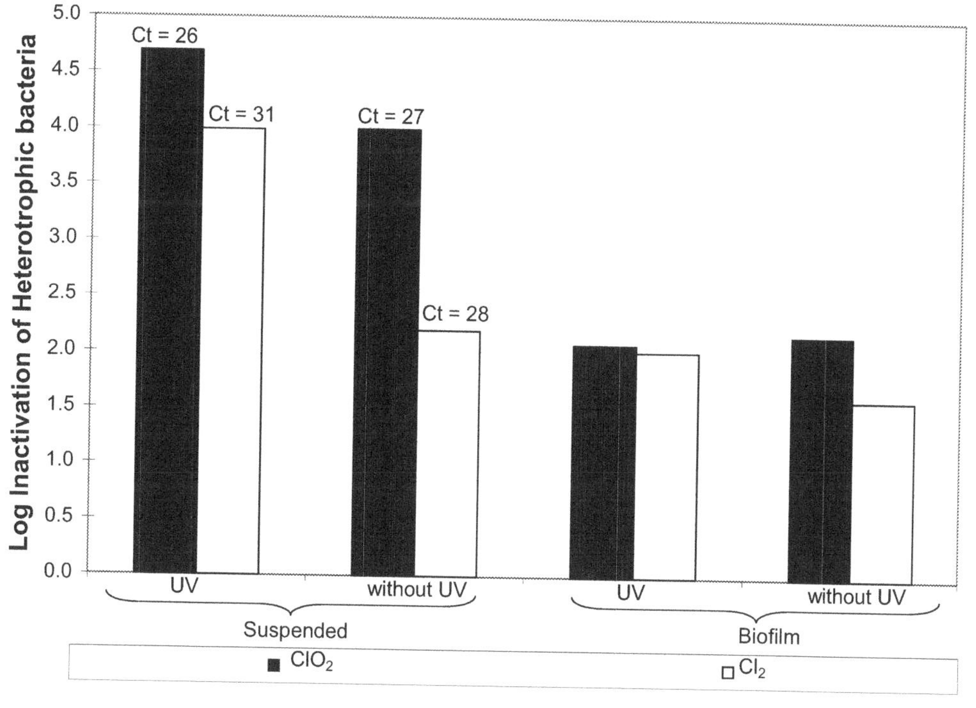

Figure 5.5 Log inactivation of heterotrophic bacteria based on equivalent Ct values (Cl_2 and ClO_2)

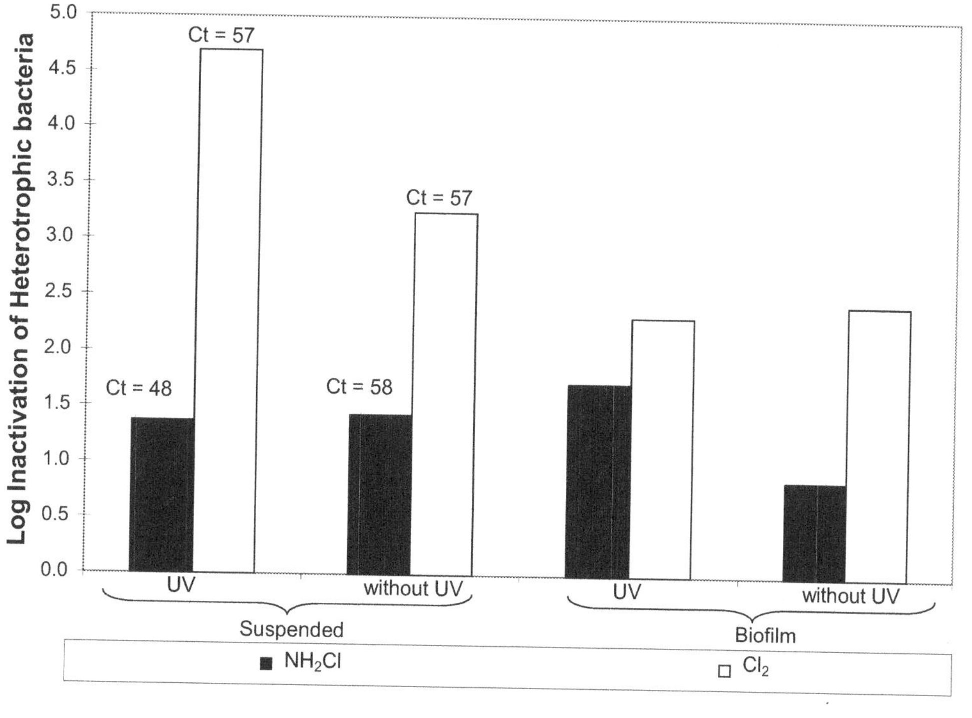

Figure 5.6 Log inactivation of heterotrophic bacteria based on equivalent Ct values (NH_2Cl and Cl_2)

Table 5.5a

P-values for the effect of chemical disinfectant on heterotrophic bacteria

	UV disinfection		No UV disinfection	
	Bulk	Biofilm	Bulk	Biofilm
Disinfection Concentration				
ClO_2	0.356	0.673	0.078	< 0.001
Cl_2	0.356	0.592	0.004	0.006
NH_2Cl	0.253	0.006	< 0.001	< 0.001

Statistical Analysis

The effect of chemical disinfectant concentration and UV disinfection on heterotrophic bacteria are provided in Tables 5.5a and 5.5b. In Tables 5.5a and 5.5b the p-values, as calculated by an ANOVA test, were used to demonstrate significance. Interestingly, the impact chemical disinfectant concentration had on suspended and biofilm. For the UV disinfected reactors the impact of disinfectant concentration was not significant for the free chlorine and chlorine dioxide ARs (Table 5.5a). However, higher disinfectant concentrations significantly increased biofilm disinfection for the chloramines AR. For ARs that did not have UV treatment, higher chemical disinfectant concentration resulted in significantly lower suspended and biofilm HPCs for free chlorine, chloramines and chlorine dioxide (biofilm HPCs only). Although the ARs with UV treatment had lower influent HPCs, it is plausible that the HPCs that were present in the influent were more vulnerable to chemical disinfectant.

At the high chemical disinfectant level, UV treatment resulted in significantly lower suspended HPCs for the three chemical disinfectants (Table 5.5b). Free chlorine and chlorine dioxide had lower bulk HPCs at the low disinfectant concentration (Table 5.5b). These data indiacted that there is a synergy between UV treatment and the chemical disinfectants, however the potential synergy depends on chemical disinfectant concentration.

Based on an equivalent Ct analysis chlorine dioxide performed better than free chlorine at reducing the suspended heterotrophic bacteria with and without the use of UV disinfection (Table 5.6). In the ClO_2 reactor (UV treated), reductions in HPCs resulted in a 0.7-log inactivation over the inactivation level in the Cl_2 reactor. In the ClO_2 reactor (untreated),

Table 5.5b

P-values for the effect of UV disinfection on heterotrophic bacteria

	Low Chemical Disinfectant		High Chemical Disinfectant	
	Bulk	Biofilm	Bulk	Biofilm
UV disinfection				
ClO_2	0.002	< 0.001	0.005	0.281
Cl_2	< 0.001	0.263	< 0.001	0.182
NH_2Cl	0.071	< 0.001	0.027	0.005

Table 5.6

P-values that compare disinfectant types on the basis of dose and equivalent Ct

	UV disinfection		No UV disinfection	
	Bulk	Biofilm	Bulk	Biofilm
Disinfection Type				
ClO_2/Cl_2 (A)	0.778	0.088	0.136	< 0.001
ClO_2/Cl_2 (B)	-	0.284	< 0.001	< 0.001
ClO_2/NH_2Cl (A)	< 0.001	0.655	0.730	< 0.001
ClO_2/NH_2Cl (B)	0.009	0.052	< 0.001	0.539
Cl_2/NH_2Cl (A)	< 0.001	0.110	< 0.001	< 0.001
Cl_2/NH_2Cl (B)	0.009	0.004	< 0.001	0.003
Equivalent Ct				
Cl_2 (A) / ClO_2 (B)	0.355	0.207	< 0.001	0.659
NH_2Cl (A) / Cl_2 (B)	< 0.001	0.215	< 0.001	< 0.001

A – Low chemical disinfectant residual concentrations.
B – High chemical disinfectant residual concentrations.

reductions in heterotrophic bacteria resulted in a 1.8-log inactivation over the inactivation level in the Cl_2 reactor. The results were shown to be significantly different ($p < 0.001$) for the reactor that received water that was not treated with UV light. ClO_2 also performed better than Cl_2 at reducing fixed heterotrophic bacteria with and without use of UV disinfection, but the results were not shown to be significant.

Cl_2 performed better than NH_2Cl at reducing the suspended heterotrophic bacteria with and without the use of UV disinfection. In the Cl_2 reactor (UV treated), reductions in HPC bacteria were 3.3-logs greater than in the NH_2Cl reactor. In the Cl_2 reactor (untreated), reductions in HPC bacteria resulted in a 1.8-log over the NH_2Cl reactor. The results were shown to be significant for both UV treated ($p=0.00004$) and untreated ($p<0.05$) waters. ClO_2 also performed better at reducing fixed heterotrophic bacteria with and without use of UV disinfection. The results were not shown to be significant. Cl_2 also performed better than NH_2Cl at reducing fixed heterotrophic bacteria with and without use of UV disinfection. The results were shown to be significant ($p<0.05$) for the reactor that received water that was not treated with UV light.

Detachment and Disinfection Kinetics

HPC results were evaluated to determine the detachment rate of the biofilm as well as the disinfection rate for each of the three disinfectants investigated. The detachment rate (k_{detach}) from the internal surface of the AR into the bulk fluid was evaluated with Equation 5.2a under steady state conditions during the acclimation period. In Equation 5.2a the growth rate of suspended cells were assumed negligible, as previous research has shown that the growth rate of suspended cells in ARs under drinking water conditions are negligible in comparison to their biofilm counterparts (Camper 1996, Jones et al. 1995). The overall disinfection rate ($k_{disinfection}$) for each disinfectant was evaluated using Equation 5.2b under the steady-state conditions established during disinfection and by incorporating an average value of k_{detach} determined from 5.2a. The k_{detach} and $k_{disinfection}$ values obtained through this analysis are provided in Table 5.7.

$$\frac{dX}{dt} = QX_i - QX + k_{detach} X_f A \quad (5.2a)$$

$$\frac{dX}{dt} = QX_i - QX + k_{detach} X_f A - k_{disin feciton} XV \quad (5.2b)$$

where Q = flow rate (mL/min)
X_i = influent heterotrophic bacteria (CFU/mL)
X = effluent heterotrophic bacteria (CFU/mL)
k_{detach} = biofilm detachment rate (min^{-1})
$k_{disinfection}$ = disinfection rate (min^{-1})
X_f = fixed heterotrophic bacteria (CFU/cm^2)
A = internal surface area of reactor
V = reactor volume

The average k_{detach} values for the ARs treated with UV disinfection during the acclimation period of the low and high disinfection stages were 1.37×10^{-4} and 2.01×10^{-4} min^{-1}, respectively. The average k_{detach} for the ARs that were treated with UV were 6.42×10^{-5} and 6.09×10^{-4} min^{-1}, respectively. The detachment rate for calculated for the acclimation period of the high chemical disinfection was an order of magnitude lower than the low disinfection phase for the reactors that were not treated with UV. For that trial, the influent water had a higher number of suspended HPCs than the other trials, which resulted in a lower detachment rate.

Table 5.7

Calculated detachment and disinfection rate coefficients

		With UV disinfection		Without UV disinfection	
		k_{detach} (min^{-1})	$k_{disinfection}$ (min^{-1})	k_{detach} (min^{-1})	$k_{disinfection}$ (min^{-1})
A	Control	1.06×10^{-4}	-	8.38×10^{-5}	-
	ClO_2	1.27×10^{-4}	27.0	1.05×10^{-4}	5.19×10^{-1}
	Cl_2	1.95×10^{-4}	28.5	3.88×10^{-5}	1.53×10^{-1}
	NH_2Cl	1.56×10^{-4}	5.41×10^{-2}	7.68×10^{-5}	8.04×10^{-2}
B	Control	1.77×10^{-4}	-	6.46×10^{-4}	-
	ClO_2	2.25×10^{-4}	19.8	6.71×10^{-4}	2.28
	Cl_2	1.57×10^{-4}	18.0	5.26×10^{-4}	1.23×10^{1}
	NH_2Cl	2.93×10^{-4}	9.41×10^{-2}	5.97×10^{-4}	4.53×10^{-1}

A – Low chemical disinfectant residual concentrations.
B – High chemical disinfectant residual concentrations.

The detachment rates determined by Camper (1996) for HPC bacteria were in the range of 0.005 to 0.15 h^{-1}. In the present investigation, the detachment rates were calculated in the range of 0.002 to 0.04 h^{-1}, which would be considered on the low end of the range reported by Camper (1996). However the influent BOM concentration in the present investigation was 100 μg/L, whereas Camper (1996) reported on influent concentrations of either 0 or 500 μg/L. Several studies have demonstrated that the detachment rate is strongly proportional to the substrate-loading rate (Gagnon 1997, Peyton and Characklis 1993, Speitel and DiGiano 1987, Stewart 1993). It is not surprising, therefore, that the detachment rate coefficients presented in Table 5.7 are in the low range of detachment rate coefficients reported by Camper (1996).

During the low chemical disinfectant stage the disinfection rates for ClO_2 and Cl_2 had a 2-log difference, between UV treated and untreated water, in suspended heterotrophic bacteria (Table 5.7). The disinfection rates for the NH_2Cl reactor were the same regardless of whether there was UV disinfection or not.

Disinfection rates for the UV treated reactors between low and high chemical disinfection residual concentrations were similar. The ClO_2 and Cl_2 reactors actually had a slightly lower disinfection rate at the high chemical dose because the HPC counts were below the detection limits during the high chemical disinfection phase. On the other hand, during the high chemical disinfectant residual stage the detachment rates were higher for the ARs that were not treated with UV.

Evaluation of Synergistic Effects

Kouame and Haas (1991) reported on the inactivation of E. coli by the combined action of free chlorine and monochloramine. They demonstrated that the interaction between free chlorine and monochloramine was synergistic. Kouame and Haas (1991) described that the process of microbial inactivation, when a combination of disinfection exists, is thought to be additive. The authors described synergism as the "combination of agents that is more effective than is expected from the single component effectiveness of its constituents". The numerical algorithm presented by Kouame and Haas (1991) can therefore by used more generally to determine if disinfectants have a synergism or beneficial combination. However, the exact chemical or cellular mechanism(s) that may lead to disinfectant sysnergism was not explained by the algorithm.

For reference purposes the algorithm is described herein. If the agents in a given combination do not interact in producing the effect observed, the following equation (Equation 5.3) is satisfied (Kouame and Haas 1991):

$$\sum_{i=1}^{n} \frac{x_i}{y_i} = 1 \qquad (5.3)$$

where x_i = concentration of the individual disinfectant in the combination
y_i = concentration of the disinfectants that individually would produce the same magnitude effect as that of the combination
i = individual disinfectant
n = number of total disinfectants

However, in the case of a synergistic interaction among disinfectants, the summation presented in Equation 5.3 would be less than 1. Similarly, in the case of antagonistic interaction among disinfectants, the summation presented in Equation 5.3 would be greater than 1.

In the present investigation, a synergistic effect between two disinfectants (i.e., UV and secondary disinfectant) was evaluated by dividing the reduction in bacterial populations that results from the combination of two disinfectant treatments (UV and secondary disinfection) by the reductions observed as a result of a single disinfection treatment (i.e., secondary disinfection), as presented in Equation 5.3. Because the same secondary disinfectant concentrations were used, with and without UV disinfection, and UV was shown to be unable to control the growth and proliferation of suspended and biofilm bacteria, a synergistic interaction could exist between UV disinfection and the secondary disinfectants. Table 5.8 shows the results of the synergistic calculations.

Relevant synergistic interaction calculations were shown to exist for low secondary disinfectant concentrations for ClO_2 and Cl_2 (suspended) and all three disinfectants (biofilm). Others showed a slightly synergistic or slightly antagonistic interaction. Interestingly, the synergistic interactions (Table 5.8) indicated in bold were previously shown to have significant differences between UV treated and untreated water, with the exception of Cl_2, with UV performing better on a log basis for biofilm reduction. In addition, the antagonistic interactions shown for NH_2Cl (high residual) were previously shown to have significant differences between UV treated and untreated water with UV performed worse on a log reduction basis. Overall, for low secondary disinfectant residuals, the results indicate that a synergistic interaction exists between UV and the secondary disinfectants. This synergic interaction was not present, or at least suppressed, for the high disinfectant residual concentrations. Although there were significant differences in the steady state HPC results, log reductions were similar, resulting in the muted synergistic effects. This could indicate that for higher concentrations of chemical disinfectants UV disinfection plays a minimized role in the control of suspended and biofilm bacterial populations.

Total Cell Counts

Mean values of AODC for suspended and biofilm cells prior to and following the addition of chemical disinfectants are shown in Tables 5.9 and 5.10 for both low and high disinfection residual concentrations. In addition, live cell counts from a live/dead cell enumeration by epifluorescence microscopy (Olympus BX-60) are provided in Tables 5.11 and 5.12, which includes the proportion of live cells. AODC results do not vary appreciably

Table 5.8

Synergistic concentration ratios ($\Sigma x_i/y_i$)

Secondary disinfectant	Suspended A	Suspended B	Biofilm A	Biofilm B
ClO_2	**0.501**	0.999	**0.510**	1.002
Cl_2	**0.483**	0.960	**0.580**	1.027
NH_2Cl	1.218	1.760	**0.524**	1.182

A – Low chemical disinfectant residual concentrations.
B – High chemical disinfection residual concentrations.

Table 5.9

Geometric mean values for suspended and biofilm total direct counts (without UV disinfection)

		No Chemical Disinfection Total Direct Counts (AODC)		Chemical Disinfection Total Direct Counts (AODC)	
		Suspended $\times 10^6$ count/mL	Biofilm $\times 10^6$ count/cm^2	Suspended $\times 10^6$ count/mL	Biofilm $\times 10^6$ count/cm^2
A	Control	0.52	0.59	0.43	1.31
	ClO_2	0.41	0.60	0.48	0.74
	Cl_2	0.51	0.60	0.45	0.68
	NH_2Cl	0.51	0.55	0.50	2.96
B	Control	1.43	8.74	0.30	6.38
	ClO_2	1.16	7.96	0.17	0.18
	Cl_2	1.24	7.79	0.17	2.67
	NH_2Cl	1.08	9.97	0.20	5.38

A – Low chemical disinfectant residual concentrations.
B – High chemical disinfectant residual concentrations.

over the entire experimental trial or between disinfectant type and dose; all are on the order of 1×10^6 counts/unit, with only slightly higher results for the biofilm cells.

The percentage of live cells decreased after chemical disinfection (Tables 5.11 and 5.12), excluding low chemical disinfection without UV disinfection. Table 5.11 reports results from experiments completed without UV disinfection. When a high disinfection residual concentration was used there was a 14-21% decrease in the proportion of live cells when compared to acclimation levels. This same trend was not observed with the low disinfection residual concentrations. Although the suspended HPCs experienced 2-logs of inactivation from control levels a similar decrease in live cells was not achieved. Table 5.12 reports results from

Table 5.10

Geometric mean values for suspended and biofilm total direct counts (with UVdisinfection)

		No Chemical Disinfection Total Direct Counts (AODC)		Chemical Disinfection Total Direct Counts (AODC)	
		Suspended $\times 10^6$ count/mL	Biofilm $\times 10^6$ count/cm^2	Suspended $\times 10^6$ count/mL	Biofilm $\times 10^6$ count/cm^2
A	Control	1.44	7.09	1.52	10.9
	ClO_2	2.66	8.10	1.14	3.77
	Cl_2	2.04	4.51	1.14	4.13
	NH_2Cl	2.53	6.12	1.14	6.82
B	Control	1.39	7.68	0.82	7.27
	ClO_2	1.50	5.88	0.45	1.93
	Cl_2	1.48	7.57	0.32	1.52
	NH_2Cl	1.23	6.50	0.59	4.52

A – Low chemical disinfectant residual concentrations.
B – High chemical disinfectant residual concentrations.

Table 5.11

Geometric mean values for suspended and biofilm live counts (without UV disinfection)

		No Chemical Disinfection		Chemical Disinfection	
		Suspended × 10^6 count/mL	Biofilm × 10^6 count/cm^2	Suspended × 10^6 count/mL	Biofilm × 10^6 count/cm^2
A	Control	0.17 (67)*	0.38 (36)	0.15 (65)	0.28 (79)
	ClO_2	-	0.33 (45)	0.083 (83)	0.26 (65)
	Cl_2	-	0.30 (50)	0.082 (82)	0.36 (47)
	NH_2Cl	-	0.27 (52)	0.11 (78)	0.35 (88)
B	Control	1.05 (73)	3.02 (35)	0.12 (40)	2.38 (37)
	ClO_2	0.67 (58)	1.38 (17)	0.04 (24)	0.11 (58)
	Cl_2	0.85 (69)	0.57 (7)	0.092 (54)	1.68 (63)
	NH_2Cl	0.71 (66)	2.68 (27)	0.09 (45)	2.30 (43)

A – Low chemical disinfectant residual concentrations.
B – High chemical disinfectant residual concentrations.
* Percentage of viable cells shown in parenthesis.

experiments completed with UV disinfection. When a low disinfection residual concentration was used there was a 14-22% decrease in the proportion of live cells when compared to acclimation levels. A similar trend was observed with the high disinfection residual concentrations.

In general, the percentage of live cells increased after chemical disinfection. The specific reasons for this are unclear, however it is known that biofilms provide some protection against changes in the source water characteristics (Percival et al. 2000) and perhaps the disinfectant is damaging the previously counted "dead" cells allowing the proportion of live cells to increase.

Table 5.12

Geometric mean values for suspended and biofilm live counts (with UV disinfection)

		No Chemical Disinfection		Chemical Disinfection	
		Suspended × 10^6 count/mL (%)	Biofilm × 10^6 count/cm^2	Suspended × 10^6 count/mL	Biofilm × 10^6 count/cm^2
A	Control	0.89 (62)*	3.60 (51)	1.07 (76)	7.07 (70)
	ClO_2	1.38 (62)	4.07 (51)	0.90 (39)	1.71 (60)
	Cl_2	1.53 (76)	2.70 (60)	0.64 (62)	2.97 (85)
	NH_2Cl	1.92 (79)	3.06 (51)	0.79 (79)	2.54 (41)
B	Control	1.22 (88)	2.40 (31)	0.61 (75)	5.31 (73)
	ClO_2	1.31 (87)	2.34 (40)	0.26 (59)	0.77 (40)
	Cl_2	1.30 (88)	2.17 (29)	0.17 (55)	0.83 (54)
	NH_2Cl	1.03 (84)	3.18 (49)	0.42 (71)	1.20 (27)

A – Low chemical disinfectant residual concentrations.
B – High chemical disinfectant residual concentrations.
* Percentage of viable cells shown in parenthesis.

Table 5.13a
P-values for the effect of chemical disinfectant on total cell counts

	UV disinfection		No UV disinfection	
	Bulk	Biofilm	Bulk	Biofilm
Disinfection Concentration				
ClO_2	0.37	0.28	0.72	0.08
Cl_2	0.03	0.09	0.52	0.036
NH_2Cl	0.20	0.17	0.71	0.46

Statistical Analysis

P-values from ANOVA tests that evaluated the effect of chemical disinfectant concentration and the effect of UV disinfection on total bacteria were performed and are reported in Tables 5.13a, 5.13b, and 5.14. Overall, results show that total bacterial counts were not significantly different ($p>0.05$) on the basis of disinfectant concentration, with the exception of chlorine in two cases (Table 5.13a). During low disinfectant concentration the total direct cell counts for water that was not treated with UV disinfection were consistently lower than any other treatment combination ($p<0.05$) for both suspended and biofilm bacteria, during acclimation and disinfection. This was not the case for high chemical disinfection concentration. During experimentation with UV, chlorine dioxide maintained lower total biofilm cells than monochloramine during low and high chemical disinfection. The same was true for chlorine during high disinfection. A similar trend was experienced without the use of UV disinfection.

Ratio of HPC/AODC in the Biofilm

The ratio of HPC (R2A plates incubated 7 days at 20°C) to AODC (total cells) was calculated at the pseudo-steady state for the different treatment conditions, to inform about the proportion of viable or cultivable bacteria in the bulk water or biofilm. Various parameters including the disinfectant, distribution system characteristics and material, or the inlet bacterial concentration, can affect the percentage of cultivable bacteria in the biofilm and bulk water.

In absence of any disinfectant, the proportions of viable bacteria were generally similar for the different ARs used in a set of experiments (low/high disinfectant dose, with/without UV), indicating that the pipes were colonized in the same way. However, these ratios could vary slightly from one set of experiments to another (Tables 5.15a and 5.15b). For instance, bulk

Table 5.13b
P-values for the effect of UV disinfection on total cell counts

	Low Chemical Disinfectant		High Chemical Disinfectant	
	Bulk	Biofilm	Bulk	Biofilm
UV disinfection				
ClO_2	0.046	0.03	0.41	0.995
Cl_2	0.011	0.04	0.53	0.09
NH_2Cl	0.003	0.25	0.20	0.40

Table 5.14
P-value comparison of disinfectants on the basis of dose and equivalent Ct for total cell counts

	UV disinfection		No UV disinfection	
	Bulk	Biofilm	Bulk	Biofilm
Disinfection Type				
ClO_2/Cl_2 (A)	0.98	0.55	0.54	0.78
ClO_2/Cl_2 (B)	0.51	0.70	0.92	0.20
ClO_2/NH_2Cl (A)	0.87	0.04	1.00	0.18
ClO_2/NH_2Cl (B)	0.99	0.02	0.92	0.002
Cl_2/NH_2Cl (A)	0.81	0.14	0.31	0.22
Cl_2/NH_2Cl (B)	0.34	0.004	1.00	0.005
Equivalent Ct				
Cl_2 (A) / ClO_2 (B)	0.31	0.16	0.49	0.21
NH_2Cl (A) / Cl_2 (B)	0.01	0.006	0.76	0.71

A – Low chemical disinfectant residual concentrations.
B – High chemical disinfectant residual concentrations.

water HPCs represented between 0.4 to 1% of the total cell concentrations for the experimental setting involving low disinfectant levels without UV treatment, while HPCs represented a few percent (from 1.5 to 9.3%) of the total cells for the three other sets of experiments. In absence of disinfectant, the HPC/AODC ratio in the biofilm was generally higher than in the bulk water. The proportion of HPC in the biofilm averaged 28% and ranged between 6.7 and 53%.

The HPC/AODC ratio decreased in the bulk water and biofilm after implementing disinfection. The proportion of cultivable bacteria decreased drastically (significant difference, $p \ll 0.05$) in the bulk water following chlorine, chlorine dioxide, and chloramine addition. In absence of disinfectant, the HPC/AODC ratios stabilized between 0.4 and 9.3%, while they ranged from 0.00032 to 0.22 % after disinfection. Regarding biofilm data, the HPC/AODC ratios were 6.7-53 % prior to disinfection, versus 0.1-6.6 % in the presence of disinfectant. The proportion of HPCs was reported for other distribution systems (Gatel et al. 1995; Block 1995; Volk et al. 2002). As a comparison, the proportion of suspended viable bacteria stabilized at 0.2 % of the total bacteria in the full-scale distribution system of Toulouse, France, after chlorine dioxide disinfection (Gatel et al. 1995). Volk et al. (2002) found the HPC/AODC ratio was lower in a Canadian system (Wiarton, Ontario): HPC represented 0.01-0.05 % of the total bacteria concentration. In another study, Block (1995) found that 1% of the bacteria (AODC measurement) was culturable on R2A agar in a pilot distribution system (pipe loop) with free chlorine. The proportions of HPCs in the bulk water containing disinfectant residuals that were observed in our study, were lower than the ones reported in the Toulouse full-scale (Gatel et al. 1995) or pilot distribution systems (Block 1995). This observation could be related to the characteristics of the model distribution system used in the present research. The presence of disinfectant residuals could be clearly observed in a polycarbonate reactor running under controlled conditions (constant water quality, no corrosion products), while the effects of disinfection could be tampered in a full-scale system, where bacterial regrowth is regulated by various parameters and their interactions. Field experiments have shown that a variety of parameters influence bacterial water quality in distribution systems (LeChevallier 1990; Camper 1996; LeChevallier et al. 1993, 1998; Volk and LeChevallier 1999). In addition to the disinfectant type and levels, other parameters including temperature, source water quality, biodegradable organic matter levels, plant operation, the conditions and composition of the distribution system, corrosion control and hydraulics, may be more influential than disinfectant for regulating bacterial regrowth and biological activity in the biofilm.

Factors Affecting the Percentage of HPC/AODC

The decrease in heterotrophic bacteria could be related to the sample type (bulk water vs. biofilm), the disinfectant type and dose, and the UV pre-treatment step (Tables 5.15a and 5.15b). The reduction in the HPC/AODC ratio was greater in bulk water samples than in the biofilm. For example, the addition of low disinfectant levels (no UV treatment) (Table 5.15a) led to a 10-100-fold decrease in the bulk water ratios, vs. a 0-15 fold decrease in the biofilm. Similarly, when higher disinfectant doses were applied, the suspended bacteria HPC/AODC ratios were decreased by a factor of 260 to 8,600, while the ratios in the biofilm showed only a 40-120-fold decrease. Such significant differences ($p < 0.05$) showed that the disinfectants were more efficient on bulk water bacteria than on biofilm bacteria. Furthermore, in most of the time, chloramines showed a lesser effect on suspended HPC bacteria (Tables 5.15a and 5.15b).

The reduction in the HPC/AODC ratio was greater at higher chemical disinfectant dosages. In absence of UV treatment (Table 5.15a), the proportion of cultivable bacteria was decreased by 10-100 folds after addition of low disinfectant residuals, compared to 260-8,600 folds, when disinfectant concentrations were doubled (significant difference, $p<0.05$). The same trend was observed with biofilm data. The proportion of cultivable bacteria in the biofilm exposed to low residuals showed reductions of 0 to 15 folds, while 40 - 120 fold decreases in the HPC/AODC ratios were observed at high residuals. Finally, the proportion of viable bacteria in bulk water samples was impacted at a higher level of magnitude when UV pre-treatment was applied. For example, the proportion of suspended HPCs was decreased by a factor of 20 to 4,000, when UV treatment was used in conjunction with low disinfectant residuals, vs. a factor of 10-100 in absence of UV oxidation. At high disinfectant concentrations, the ratio could be decreased by a factor up to 33,000 in presence of UV and chlorine dioxide, while it decreased by 260-8,600 folds without UV.

Table 5.15a
Percentage of HPC/AODC without UV disinfection

		No Chemical Disinfection HPC/AODC (%)		Chemical Disinfection HPC/AODC (%)	
		Bulk water	Biofilm	Bulk water	Biofilm
A	Control	0.8	32	1.4	6.6
	ClO_2	1.0	39	0.1	6.6
	Cl_2	0.4	13	0.01	0.4
	NH_2Cl	0.4	25	0.09	1.0
B	Control	3.8	11	50	16
	ClO_2	1.6	6.7	0.006	0.4
	Cl_2	1.5	11	0.05	0.1
	NH_2Cl	3.7	9.7	0.2	0.1

A – Low chemical disinfectant residual concentrations
B – High chemical disinfectant residual concentrations

Table 5.15b

Percentage of HPC/AODC with UV disinfection

		No Chemical Disinfection HPC/AODC (%)		Chemical Disinfection HPC/AODC (%)	
		Bulk water	Biofilm	Bulk water	Biofilm
A	Control	7.3	46	1.4	6.2
	ClO_2	9.3	61	< 0.001	0.5
	Cl_2	5.1	32	<0.001	0.1
	NH_2Cl	4.6	53	0.07	0.1
B	Control	3.4	27	6.4	13
	ClO_2	2.3	26	<0.001	0.6
	Cl_2	2.7	25	<0.001	0.5
	NH_2Cl	2.4	27	0.22	1.0

A – Low chemical disinfectant residual concentrations.
B – High chemical disinfectant residual concentrations.

Chemical Analysis

Total Organic Carbon

Average results for TOC analysis are shown in Table 5.16. There were no data available for low chemical disinfection experimentation due to analytical issues. Average TOC results for high chemical disinfection dose indicates that when no chemical disinfection is applied there was an equal utilization of organic carbon (0.03 mg/L) within all 4 reactors, which was expected, as they are replicates of each other. When secondary disinfection was initiated the organic carbon utilization dropped off in the reactors receiving chemical disinfectant, while there was observed utilization within the control reactor (0.04 mg/L).

Disinfection By-Products

Average results for DBPs are shown in Table 5.17. There were no data available for experimentation without the use of UV disinfection or for the low disinfection residual concentration during the acclimation portion. During the low disinfection stage HAA results were the highest for the reactor that was disinfected with chlorine, followed by monochloramine and chlorine dioxide. This result is expected as chlorine has the highest DBP potential of the three disinfectants (Droste 1997). Results from TTHM analysis resulted in very little change between reactors with the majority of samples were below detection limits.

During the high disinfection stage HAA results followed a similar trend as the low disinfection stage; results were the highest for the reactor that was disinfected with chlorine, followed by monochloramine and chlorine dioxide. Results from TTHM analysis resulted in very little change between reactors with results in the range of 25 μg/L. During this stage of experimentation, the source water (BAC) experienced a dramatic increase in TTHM formation from 0 μg/L during acclimation to 40 μg/L during disinfection. At this time it was determined that the GAC filter bed had been exhausted and the media was replaced.

Table 5.16
Average TOC results (with UV disinfection)

		No Chemical Disinfection	Chemical Disinfection
		TOC (mg/L)	TOC (mg/L)
A	Influent	-	-
	Control	-	-
	ClO_2	-	-
	Cl_2	-	-
	NH_2Cl	-	-
B	Influent	0.12	0.23
	Control	0.09	0.19
	ClO_2	0.10	0.26
	Cl_2	0.10	0.23
	NH_2Cl	0.09	0.23

A – Low chemical disinfectant residual concentrations.
B – High chemical disinfectant residual concentrations.

Table 5.17
Average disinfection by-product results (with UV disinfection)

		No Chemical Disinfection		Chemical Disinfection	
		HAA_6 µg/L	TTHM µg/L	HAA_6 µg/L	TTHM µg/L
A	Control	n.d.	n.d.	0.4	0
	ClO_2	n.d.	n.d.	10	0
	Cl_2	n.d.	n.d.	19	0.2
	NH_2Cl	n.d.	n.d.	13	0.9
B	Control	0	12	0.1	25
	ClO_2	0.6	7	2	24
	Cl_2	n.d.	7	11	26
	NH_2Cl	0	14	6	25

A – Low chemical disinfectant residual concentrations.
B – High chemical disinfectant residual concentrations.
n.d. – no data.

SUMMARY

This study evaluated the synergistic effects of UV disinfection coupled with secondary disinfectants using a model distribution system. Results indicate that there are synergistic effects between UV disinfection and the chemical disinfectants for controlling microbiological regrowth in drinking water distribution systems. Chlorine dioxide (ClO_2) performed better than free chlorine (Cl_2) on an equivalent Ct basis whether or not UV disinfection was employed. However, in the reactors fed with UV treated water both chlorine dioxide (ClO_2) and free chlorine (Cl_2) performed better at reducing the suspended heterotrophic bacteria. Disinfecting the source water with UV light appears to impact the suspended bacteria by increasing vulnerability to chemical disinfection. However, the established biofilm maintains its ecological properties and results in similar behavior to chemical disinfection. The use of UV without a secondary disinfectant is unable to control the proliferation of suspended or biofilm bacteria. It is hypothesized that cellular mutations resulting from UV disinfection cause the suspended cells to be more susceptible to chemical disinfection. The HPC results were evaluated to determine the detachment rate of the biofilm from the internal surface of the reactors as well as the disinfection coefficient for each disinfectant investigated. The detachment rates were calculated in the range of 0.002 to 0.04 h^{-1} and were reasonably consistent in the presence and absence of UV disinfection. Overall, these findings indicate that UV prior to chemical disinfection enhances microbial control in distribution systems. Practically, these results would indicate that a dual-barrier disinfection system should be considered for pathogen protection at the plant and distribution system.

CHAPTER 6
IMPACT OF UV DISINFECTION AND NOM ON BIOFILM REGROWTH WITH PRE-CHLORINATION (PHASE 2)

INTRODUCTION

Natural organic matter is present in untreated surface and ground water, sometimes in large concentrations. In many cases, particularly in rural communities, water is used directly from the source (in the case of well water) or it is chlorinated and distributed to consumers (small-scale treatment systems). In order to protect the consumer from potentially pathogenic bacteria and limit the production of DBPs some communities and households are turning to UV disinfection as a treatment option. The problem arises when the organic material present is at high enough concentrations to both protects bacteria from UV disinfection and act as a substrate for microbial growth. Vallino et al. (1996) found that simulations indicate bacterial growth rate and yield are strongly correlated to the oxidation state of the labile DOM and that bacterial growth cannot always be explained by a single constraint (such as C:N ratio), since several constraints are often active simultaneously and continuously change with environmental conditions.

The overall objective of phase 2 was to evaluate the impact of UV disinfection and NOM on biofilm accumulation in drinking water distribution systems with pre-chlorination. In addition, the research compared the effectiveness of chlorine dioxide, free chlorine, and monochloramine as secondary disinfectants for reducing bacterial regrowth (suspended and biofilm cells) in the presence and absence of UV disinfection. Previous studies have evaluated free chlorine and monochloramine for controlling biofilm growth, however there is a general lack of published data concerning chlorine dioxide in this area. The use of UV disinfection at the start of drinking water treatment plants can act to both control the microbial population entering the plant and possibly act to oxidize organic material. It is important to understand what impact UV disinfection has on organic matter present in the source water and what the impact means for distribution water quality and biostability. Natural organic matter has the potential to absorb and scatter UV light, thus decreasing UV intensity and protecting microbial populations. In cases where UV light is the only barrier against microbiological outbreaks the impact that NOM has on UV efficacy is very important.

MATERIALS AND METHODS

Methods for the operation of the UV disinfection assembly and the annular reactors are described in detail in Chapter 3. Figure 6.1 illustrates the specific details related to the operation of the process water being passed through UV disinfection system and subsequently pumped into the reactors. Source water was prepared by pumping tap water through two GAC columns followed by subsequent UV disinfection, as previously described. In addition to carbon-filtered tap water, commercially available humic material (Aldrich humic acid) and hypochlorous acid (HOCl) were pumped into the UV reactor. The annular reactors were augmented with nitrogen and phosphorous compounds as in the previous phase to ensure a carbon limiting system.

Secondary disinfection commenced after the 4-week acclimation period was completed and a steady-state biofilm was achieved, as indicated by a relatively constant number of biofilm

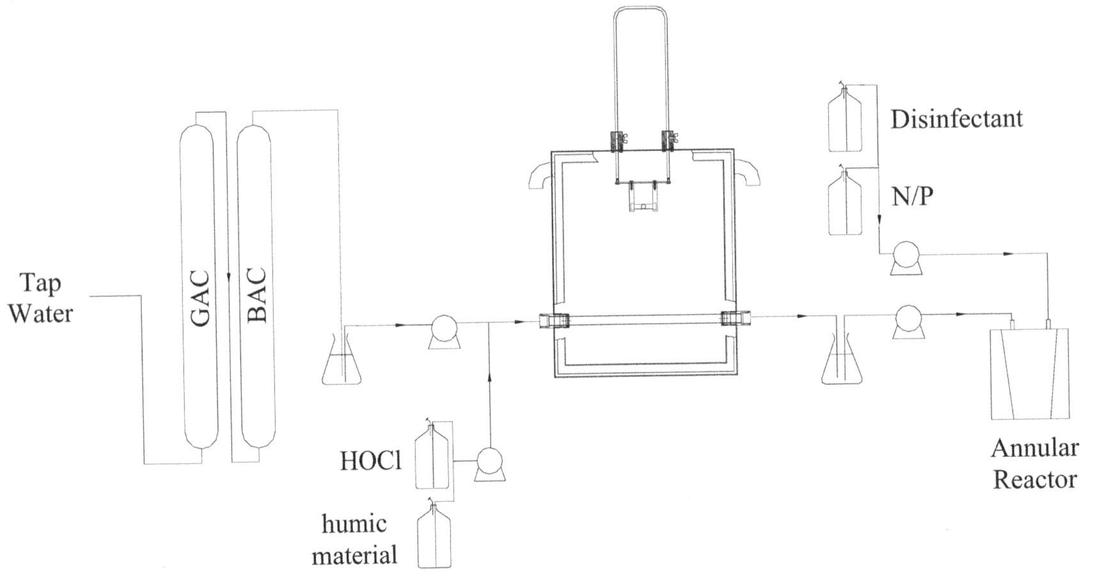

Figure 6.1 Schematic of bench-scale system during Phase 2

heterotrophic bacteria. Data collected during the subsequent disinfection period enabled a direct comparison of chlorine dioxide, free chlorine, and chloramines for controlling biofilm growth in distribution systems. One of the four ARs did not receive any secondary disinfectant residual, as shown in the experimental matrix (Table 6.1). Experimental trials were performed with and without a pre-chlorination dose applied prior to UV disinfection. The pre-chlorination disinfectant concentration goal was 1 mg/L in the UV clearwell.

Biofilm Structure

Coupon Preparation for Transport

Polycarbonate coupons, for microscopic analysis, were removed aseptically from the annular reactor and transferred to a wetted, autoclaved sponge that had been prepared to accommodate the coupon and secure it in place. The sponge and coupon were then transferred into a sterile stomacher bag (177 × 304 mm) for transport. The removed coupons were replaced with cleaned, sterilized coupons to maintain constant volume and hydraulic conditions.

Table 6.1

Experimental matrix for Phase 2

Pre-Chlorination Dose (mg/L)	Disinfectant Residual Concentration (mg/L)			
	Control: No Disinfectant	Chlorine Dioxide	Free Chlorine	Monochloramine
0.0	0.0	0.25	0.5	1.0
1.0	0.0	0.25	0.5	1.0

Coupon Preparation for Microscopic Analysis

Microscopic analysis was completed at Indiana University Kokomo. Coupons were stained at 21°C by immersion in a solution of 0.01% (w/v) acridine orange (ICN Biochemicals) for 5 minutes. After staining, the coupons were gently rinsed in sterile deionized water and dried. The coupons were observed by epifluorescence microscopy using an Olympus BX-60 microscope with a blue (450-480 nm) excitation filter used in combination with a 520-nm barrier filter. Digital pictures were taken using a CCD camera and an image analysis software package (Esprit™ version 1.0).

Bacterial Identification

Bacterial isolates from samples collected during phase 2 were identified. The most predominant colonies from R2A plates (heterotrophic plate counts from various samples) were isolated on several occasions and maintained on R2A slants. These heterotrophic bacteria were identified by Gram staining, the oxidase test, and API identification systems (API 20E and API 20NE from BioMérieux, Hazelwood, MO). These two systems are specific for Gram negative bacteria.

RESULTS AND DISCUSSION

Microbial Analysis

Heterotrophic Bacteria

For experimentation without pre-chlorination, the concentration of suspended heterotrophic bacterial cells (post-BAC column) during the acclimation period had a geometric mean of 780 CFU/mL. Influent suspended bacterial cell concentrations (post-UV treatment) during the acclimation period had a geometric mean of 2500 CFU/mL. Finally, the concentration of bacteria during the acclimation period in the control AR had a geometric mean of 4.77×10^4 CFU/mL.

For pre-chlorination experimentation, the concentration of suspended heterotrophic bacterial cells (post-BAC column) during the acclimation period had a geometric mean of 450 CFU/mL. Influent suspended bacterial cell concentrations (post-UV treatment) during the acclimation period had a geometric mean of 11 CFU/mL, but in most cases was <1 CFU/mL. Finally, the concentration of bacteria during the acclimation period in the control AR had a geometric mean of 7.27×10^2 CFU/mL.

Figure 6.2 illustrates the result of using chlorine in combination with UV on heterotrophic bacteria. Heterotrophic bacteria in the UV clearwell reached a concentration of approximately 1×10^5 CFU/mL without pre-chlorination. When pre-chlorination was used the heterotrophic bacterial concentration was consistently less than 1 CFU/mL. Suspended and biofilm heterotrophic plate counts (HPCs) for the ARs that received no pre-chlorination are illustrated in Figures 6.3 and 6.5, respectively.

Suspended and biofilm heterotrophic plate counts (HPCs) for the ARs that received a pre-chlorination dose are illustrated in Figures 6.4 and 6.6, respectively. Disinfection in the ARs began after the fourth week of experimentation when a steady state biofilm was established,

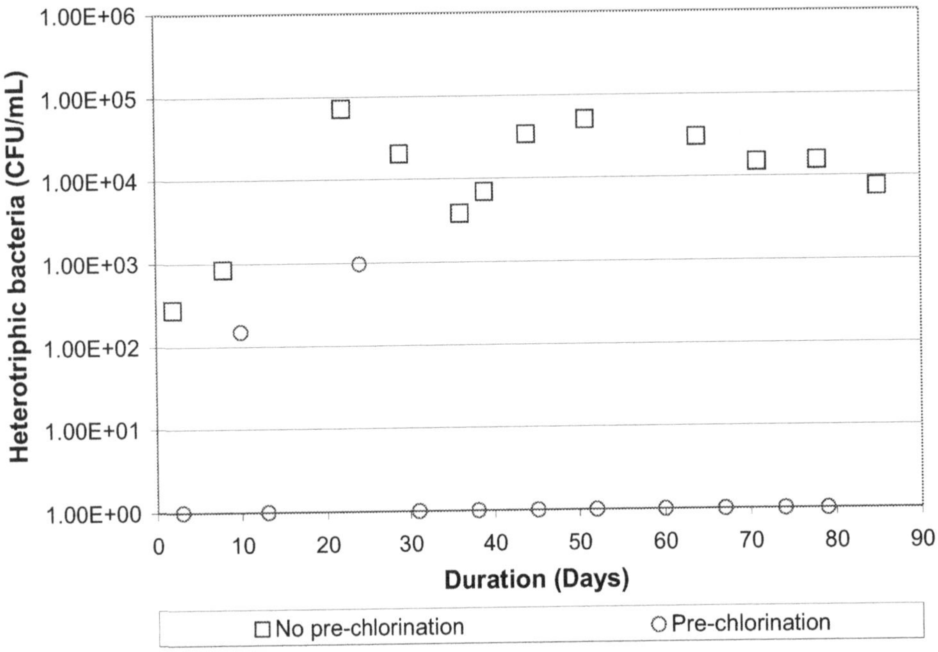

Figure 6.2 Suspeneded HPCs in UV clearwell (with and without pre-chlorination)

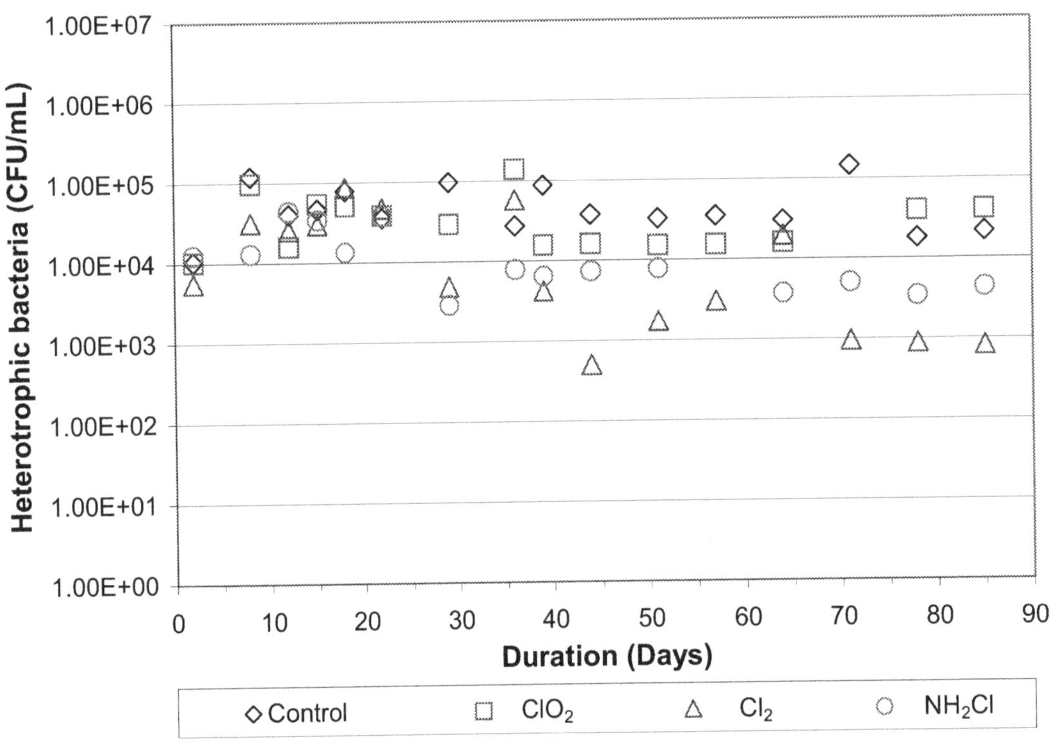

Figure 6.3 Suspended heterotrophic bacteria in ARs (no pre-chlorination), with secondary disinfection applied on day 30

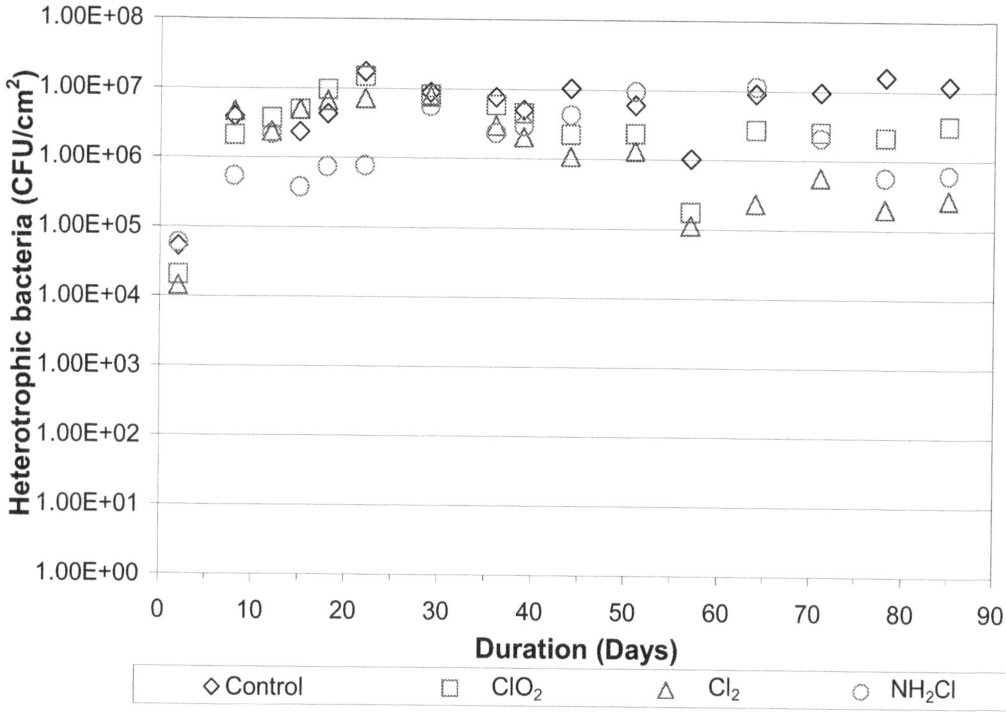

Figure 6.4 Biofilm heterotrophic bacteria in ARs (pre-chlorination), with secondary disinfection applied on day 30

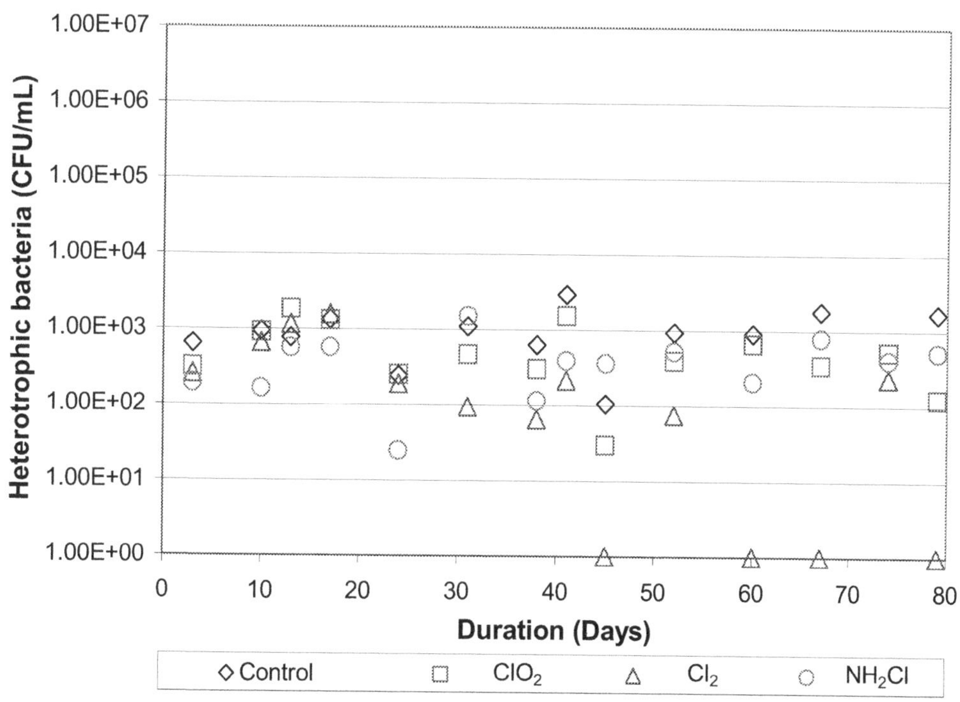

Figure 6.5 Suspended heterotrophic bacteria in ARs (no pre-chlorination), with secondary disinfection applied on day 30

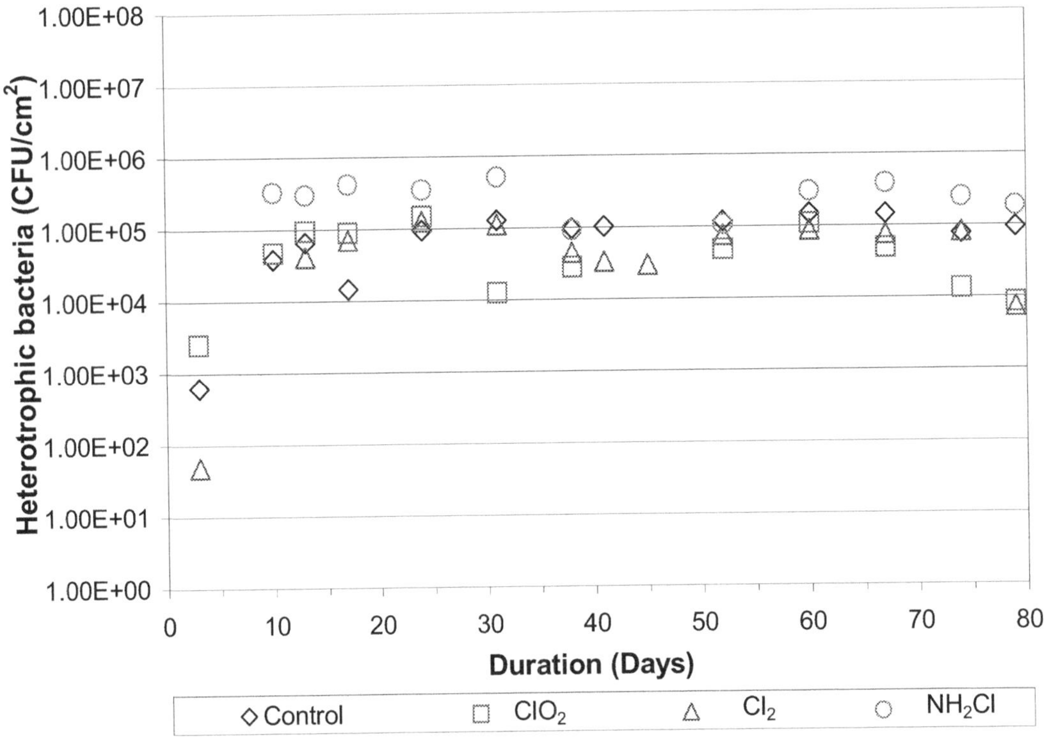

Figure 6.6 Biofilm heterotrophic bacteria in ARs (pre-chlorination), with secondary disinfection applied on day 30

which is consistent with biofilm work of others (Camper 1996). Throughout experimentation the pH of the water was 7.4 ± 0.4 and the temperature was 20 ± 1 °C.

Steady state geometric mean values for suspended and biofilm HPCs prior to chemical disinfection are shown in Table 6.2. Geometric mean values for suspended and biofilm HPCs reflect the last four sample points just prior to disinfection and the last four sample points of the experimental trial when the bacterial concentrations had reached pseudo-steady state. Log reductions in heterotrophic bacteria for each of the disinfectants are also shown in Table 6.2.

Table 6.3a shows the average disinfectant concentrations in the reactor effluents and the calculated contact time (Ct) for the ARs without pre-chlorination. Table 6.3b shows the average disinfectant concentrations in the UV clearwell and in the reactor effluents with pre-chlorination. Calculated Ct is also reported for each location.

Disinfection efficacy for each component (with and without pre-chlorination) was made on the basis of log reduction data presented in Table 6.2. Figure 6.7 graphically illustrates the comparison. Experimental trials with the use of pre-chlorination experienced the complete inactivation of suspended heterotrophic bacteria in the reactor disinfected with chlorine. Suspended HPCs experienced 1.3-logs of inactivation without pre-chlorination and 2.6-logs of inactivation with pre-chlorination during secondary disinfection. This may have occurred as a result of the decreased number of influent and biofilm bacteria and the combination of pre-chlorination and secondary disinfection concentration, which gave an average free chlorine concentration of 0.96 mg/L during disinfection.

Table 6.2

Geometric mean values for suspended and biofilm heterotrophic bacteria

		No Chemical Disinfection HPCs		Chemical Disinfection HPCs		Log Reduction in HPCs	
		Suspended × 10^4 CFU/mL	Biofilm × 10^6 CFU/cm^2	Suspended × 10^4 CFU/mL	Biofilm × 10^6 CFU/cm^2	Suspended	Biofilm
A	Control	4.77	5.64	3.62	11.3	0.12	-0.30
	ClO_2	3.62	7.19	2.96	2.62	0.09	0.63
	Cl_2	4.19	4.88	0.19	0.29	1.28	1.59
	NH_2Cl	2.95	0.83	0.41	1.69	0.95	0.82
B	Control	0.073	0.058	0.14	0.102	-0.28	-0.24
	ClO_2	0.074	0.063	0.035	0.017	0.60	0.76
	Cl_2	0.043	0.079	0.0004	0.036	2.55	0.46
	NH_2Cl	0.033	0.39	0.04	0.26	0.51	-0.41

A – No pre-chlorination.
B – Pre-chlorination.

Table 6.3a

Chemical disinfectant residual concentrations and Ct values (no pre-chlorination)

	Target Residual Concentration (mg/L)	Average Measured Residual Concentration (Standard Deviation)	Ct[1] (mg/L•min)
Control	0.0	-	
ClO_2	0.25	0.11 (0.04)	7
Cl_2	0.5	0.44 (0.20)	26
NH_2Cl	1.0	0.93 (0.28)	56

Table 6.3b

Chemical disinfectant residual concentrations and Ct values (pre-chlorination)

		Target Residual Concentration (mg/L)	Average Measured Residual Concentration (Standard Deviation)	Ct[1] (mg/L•min)
Acclimation	UV Cl_2	1.0	0.99 (0.12)	10
	AR – Cl_2	-	0.45 (0.13)	27
Disinfection	UV Cl_2	1.0	0.95 (0.11)	10
	Control – Cl_2[2]	-	0.22 (0.14)	13
	ClO_2 – Cl_2	-	0.23 (0.15)	14
	ClO_2	0.25	0.18 (0.07)	11
	Cl_2	0.5	0.96 (0.06)	58
	NH_2Cl – Cl_2	-	0.10 (0.02)	6
	NH_2Cl	1.0	1.24 (0.41)	74

1 – Ct calculated from the average measured chemical residual concentration.
2 – Control-Cl_2 indicates that the control reactor was prechlorinated.

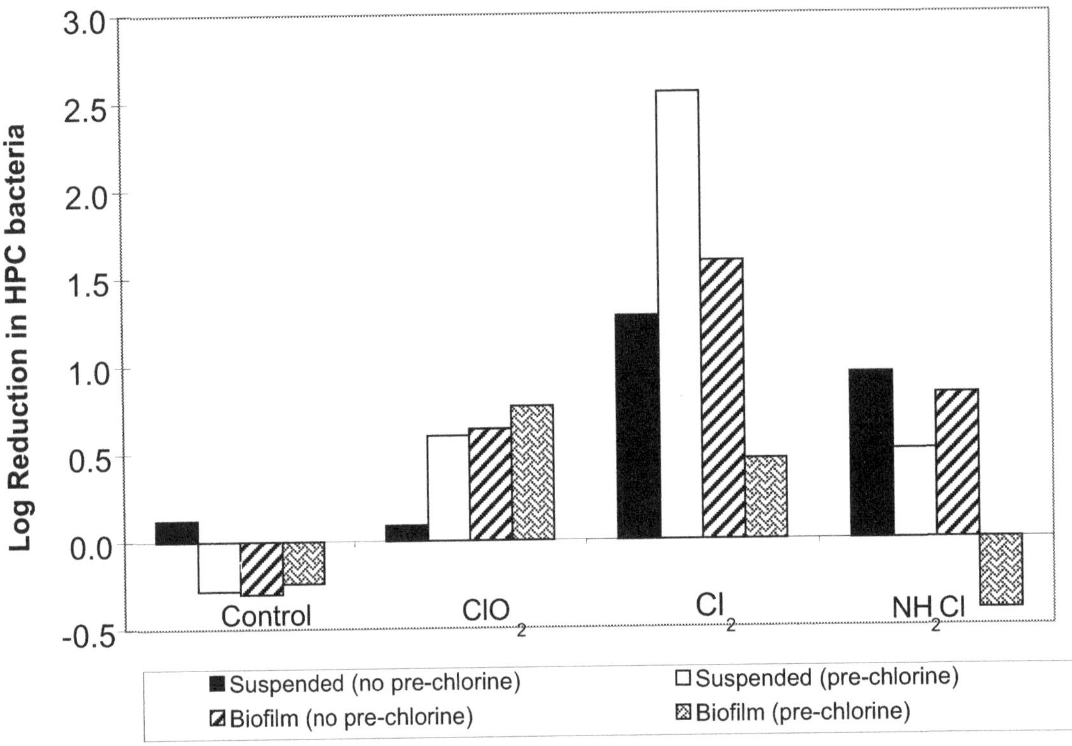

Figure 6.7 Log inactivation of heterotrophic bacteria (with and without pre-chlorination)

The inactivation values with chlorine dioxide was similarly affected by pre-chlorination as the inactivation of suspended bacteria increased 0.5-log over levels observed when pre-chlorination was not used. The application of monochloramine did not see result in trends as the log inactivation actually decreased. However, all three disinfectants were able to reduce suspended HPCs below levels observed when pre-chlorination was not used.

Overall, biofilm results displayed opposite trends as the log reductions decreased in all cases with both monochloramine and chlorine. This could be the result of acclimation biofilm levels during pre-chlorination being the same as disinfection biofilm levels when pre-chlorination was not used. The small reductions observed may be the result of biofilm bacteria having been selected to be chlorine resistant which were not significantly affected by additional chlorine disinfection. However, the reactor disinfected with ClO_2 experienced a slight increase in HPC inactivation, indicating an increased ability for biofilm disinfection.

Statistical Analysis

P-values from ANOVA tests that evaluated the effect of pre-chlorination on heterotrophic bacteria are provided in Table 6.4. Chlorination prior to UV disinfection had a significant impact on both suspended and biofilm HPCs, with the exception of chloramines on biofilm HPCs (Table 6.4). This was due to the chloramines reactor having slightly lower biofilm levels without pre-chlorination (Figure 6.5) and slightly higher biofilm levels with pre-chlorination

Table 6.4

P-values for the effect of pre-chlorination on heterotrophic bacteria

	Acclimation		Disinfection	
	Bulk	Biofilm	Bulk	Biofilm
Pre-chlorination				
Control	0.002	<0.001	0.017	-
ClO_2	0.004	0.015	<0.001	<0.001
Cl_2	0.015	<0.001	<0.001	0.024
NH_2Cl	0.003	0.15	<0.001	0.24

(Figure 6.6). The same trend was observed during disinfection. In fact, heterotrophic bacteria did not change between acclimation and disinfection (p=0.22-0.45) in the chloramines reactors. Acclimation suspended HPC levels during pre-chlorination were statistically the same as disinfection suspended HPC levels without pre-chlorination (p=0.34). With prechlorination, suspended HPC levels declined again following the addition of secondary disinfection in the reactor. Perhaps the increased disinfection level was due to the staged application of free chlorine. The same trend was observed for biofilm HPCs (p=0.71).

Chlorine performed better than NH_2Cl and ClO_2 at reducing suspended heterotrophic bacteria with and without pre-chlorination (Table 6.5). In the Cl_2 reactor (without pre-chlorination), reductions in suspended HPCs were 0.33-logs over the NH_2Cl reactor and 1.19-logs over the ClO_2 reactor (Table 6.2). With pre-chlorination, reductions in suspended HPCs in the Cl_2 reactor were 2-logs over the NH_2Cl reactor and 1.95-logs over the ClO_2 reactor. The results were shown to be significant with and without pre-chlorination.

Detachment and Disinfection Kinetics

HPC results were evaluated to determine the detachment rate of the biofilm as well as the disinfection rate for each of the three disinfectants investigated for experimentation without pre-chlorination. Detachment and disinfection rates were not determined for experimentation with pre-chlorination as detachment and disinfection occurred simultaneously within the reactor during acclimation. The detachment rate (k_{detach}) from the internal surface of the AR into the bulk fluid and the overall disinfection rate ($k_{disinfection}$) for each disinfectant was evaluated using the same rationale used in Chapter 5. The k_{detach} and $k_{disinfection}$ values obtained through this analysis are provided in Table 6.6. The average values of k_{detach} for the ARs treated with UV disinfection was 4.01×10^{-5} min^{-1} during the acclimation period.

The detachment rates determined by Camper (1996) for HPC bacteria were in the range of 0.005 to 0.15 h^{-1}. In the present investigation, the detachment rates were calculated in the range of 0.001 to 0.01 h^{-1}, which would be considered on the low end of the range reported by Camper (1996). Several studies have demonstrated that the detachment rate is strongly proportional to the substrate-loading rate (Gagnon 1997, Peyton and Characklis 1993, Speitel and DiGiano 1987, Stewart 1993). It is not surprising, therefore, that the detachment rate coefficients presented in Table 6.6 are in the low range of detachment rate coefficients reported by Camper (1996). It is also not surprising that the detachment rate coefficients are lower than

Table 6.5

P-values for the effect of disinfectant type on heterotrophic bacteria

	No pre-chlorination		Pre-chlorination	
	Bulk	Biofilm	Bulk	Biofilm
Disinfection Type				
ClO_2/Cl_2	0.017	<0.001	0.034	0.56
ClO_2/NH_2Cl	0.024	0.71	0.71	0.003
Cl_2/NH_2Cl	<0.001	0.24	0.020	0.003

those observed in the previous phase as the suspended HPC acclimation values are statistically the same (p=0.91) regardless of the carbon source. The biofilm HPC values on the other hand are significantly higher (1-log) for the Aldrich humic acid than for the BOM (p=0.001).

Total Bacterial Cells

Mean values for suspended and biofilm AODC prior to and following the addition of secondary chemical disinfectants are shown in Table 6.7. In addition, live cell counts from a live/dead cell enumeration by epifluorescence microscopy (Olympus BX-60) are provided in Table 6.8, which includes the proportion of live cells.

For experimentation without pre-chlorination, the percentage of live cells (i.e., live cells / total direct cell counts) increased after chemical disinfection for both suspended and biofilm bacteria, as did the number of total biofilm cells (p<0.001) without any significant change, from acclimation levels, in the number of total suspended cells (p=0.45). The number of live cells also experienced a significant increase in suspended (p=0.02) and biofilm cells (p=0.003). For experimentation with pre-chlorination the proportion of live cells decreased after secondary chemical disinfection for both suspended and biofilm bacteria. There was no significant change, from acclimation levels, in the number of total suspended (p=0.92) and biofilm (p=0.57) cells, or the number of live suspended (p=0.86) and biofilm cells (p=0.15).

Statistical Analysis

P-values from ANOVA tests that evaluated the effect of pre-chlorination on total and live cells are provided in Tables 6.9 and 6.10, respectively. In general chlorination prior to UV disinfection did not have a significant impact on suspended total or live cells. Pre-chlorination did have a significant impact on biofilm total and live cells, following secondary disinfection. The presence of humic material without pre-chlorination allowed for a larger biofilm to establish and increased the number of total and live bacteria present during disinfection. The presence of the denser biofilm may have provided improved protection against chemical disinfection.

Total biofilm cells were significantly lower with pre-chlorination during disinfection but were statistically the same during acclimation. The same general trend was observed for live suspended and biofilm cells. During disinfection the number of live cells in the bulk fluid increased significantly (p=0.02) as did the number of live cells in the biofilm (p=0.003). In contrast, disinfectant type did not have an impact on total or live cells (Table 6.11 and 6.12).

Table 6.6

Calculated detachment and disinfection rate coefficients

	Without pre-chlorination	
	k_{detach} (min^{-1})	$k_{disinfection}$ (min^{-1})
Control	4.31×10^{-5}	-
ClO_2	2.35×10^{-5}	-1.76×10^{-3}
Cl_2	4.22×10^{-5}	8.02×10^{-2}
NH_2Cl	1.52×10^{-4}	6.27×10^{-2}

Table 6.7

Geometric mean values for suspended and biofilm AODC

		No Chemical Disinfection AODC		Chemical Disinfection AODC	
		Suspended $\times 10^6$ count/mL	Biofilm $\times 10^6$ count/cm^2	Suspended $\times 10^6$ count/mL	Biofilm $\times 10^6$ count/cm^2
A	Control	1.45	7.06	2.54	202
	ClO_2	1.48	5.21	2.59	97.6
	Cl_2	1.27	5.35	2.81	105
	NH_2Cl	1.20	6.09	3.13	85.8
B	Control	2.38	3.54	1.87	5.74
	ClO_2	2.31	1.84	2.42	4.02
	Cl_2	3.02	3.25	1.90	3.08
	NH_2Cl	2.63	3.78	2.33	5.11

A – No pre-chlorination.
B – Pre-chlorination.

Table 6.8

Geometric mean values for suspended and biofilm live counts

		No Chemical Disinfection Live Counts		Chemical Disinfection Live Counts	
		Suspended $\times 10^6$ count/mL (%)	Biofilm $\times 10^6$ count/cm^2 (%)	Suspended $\times 10^6$ count/mL (%)	Biofilm $\times 10^6$ count/cm^2 (%)
A	Control	0.47 (33)	3.94 (56)	1.84 (78)	125 (62)
	ClO_2	0.72 (49)	2.11 (41)	2.10 (77)	62.9 (61)
	Cl_2	0.33 (26)	1.99 (37)	1.90 (69)	94.7 (79)
	NH_2Cl	0.48 (40)	2.72 (45)	2.25 (81)	62.7 (58)
B	Control	1.75 (81)	2.89 (74)	1.14 (58)	1.90 (51)
	ClO_2	1.53 (71)	1.76 (84)	0.97 (48)	1.99 (53)
	Cl_2	3.23 (88)	2.58 (73)	0.92 (49)	1.52 (46)
	NH_2Cl	2.19 (86)	2.86 (79)	1.17 (51)	2.67 (54)

A – No pre-chlorination.
B – Pre-chlorination.

Table 6.9

P-values for the effect of pre-chlorination on total bacteria

	Acclimation		Disinfection	
	Bulk	Biofilm	Bulk	Biofilm
Pre-chlorination				
Control	0.96	0.027		
ClO_2	0.29	0.12	0.44	0.004
Cl_2	0.013	0.42	0.11	<0.001
NH_2Cl	0.12	0.37	0.45	0.012

Table 6.10

P-values for the effect of pre-chlorination on live bacteria

	Acclimation		Disinfection	
	Bulk	Biofilm	Bulk	Biofilm
Pre-chlorination				
Control	0.56	0.04		
ClO_2	0.18	0.80	0.04	0.01
Cl_2	0.003	0.37	0.09	0.002
NH_2Cl	0.03	0.66	0.10	0.07

Table 6.11

P-values that compare secondary disinfectants on total bacteria

	No pre-chlorination		Pre-chlorination	
	Bulk	Biofilm	Bulk	Biofilm
Disinfection Type				
ClO_2/Cl_2	0.97	0.78	0.63	0.76
ClO_2/NH_2Cl	0.89	0.95	0.59	0.22
Cl_2/NH_2Cl	0.87	0.86	0.25	0.20

Table 6.12

P-values that compare secondary disinfectants on live bacteria

	No pre-chlorination		Pre-chlorination	
	Bulk	Biofilm	Bulk	Biofilm
Disinfection Type				
ClO_2/Cl_2	0.80	0.44	0.79	0.59
ClO_2/NH_2Cl	0.85	0.90	0.56	0.37
Cl_2/NH_2Cl	0.67	0.70	0.42	0.19

Qualitative Analysis

Experimentation without Pre-chlorination. Reactor coupons were analysed microscopically to investigate the structure of the biofilm in the control reactor and after exposure to the three secondary disinfectants investigated (Figures 6.8 and 6.9). The structure of the biofilm in the control reactor is very dense as a result of the humic material used as the organic source (Figure 6.8). Analysis of the coupons found primarily rod-shaped bacteria, with some filamentous bacteria. There is an obvious difference, due to disinfection, in the structure of the biofilm and the number of total bacteria present as determined from the acridine orange stained used (Figure 6.9).

Figure 6.10 illustrates the fouling of the coupons caused by the humic material and the result of oxidation with chlorine dioxide. The control coupons removed from the reactor were completely fouled once acclimation HPC values reached steady state. Oxidation with chlorine and chlorine dioxide created "clear" patches on the coupons where the biofilm had been removed. Coupons removed from the chlorine reactor appeared to have been completely oxidized, as the brown colour was not present, while coupons removed from the chloramine reactor did not appear different from the control coupon.

Experimentation with Pre-chlorination. Reactor coupons at steady-state were analysed microscopically to investigate the structure of the biofilm in the control reactor and after exposure to the three secondary disinfectants investigated (Figure 6.11). There was an obvious difference, due to disinfection, in the structure of the biofilm and the number of total bacteria present as determined from the acridine orange stained used. Similarly there was also a difference in the biofilm structure as a result of pre-chlorination. Coupons removed from the reactors did not exhibit the brown colour that was observed previously indicating that the humic material had been oxidized prior to entering the reactor and while in the reactor. There was also a reduced opportunity for a biofilm to form due to the presence of free-chlorine throughout the entire experimental trial.

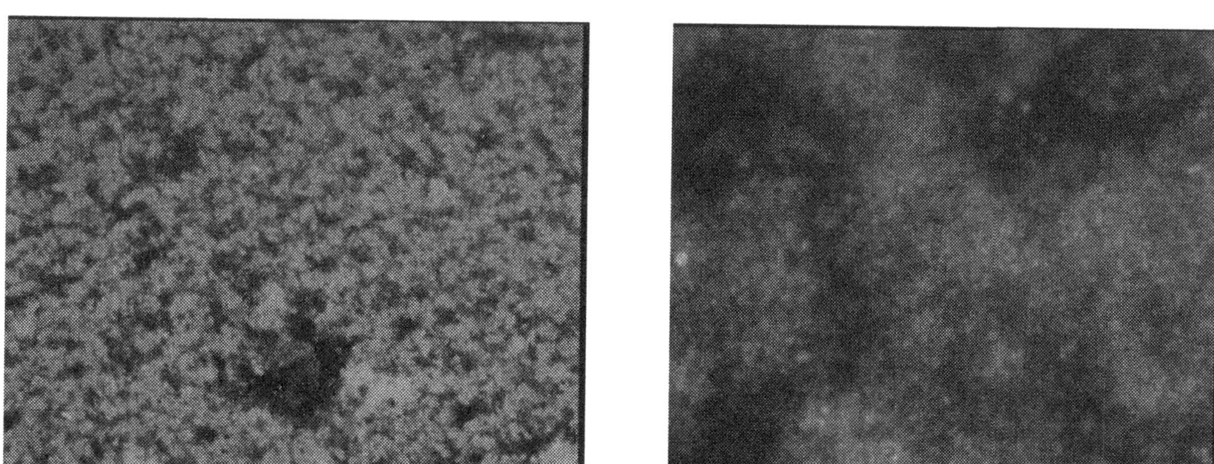

Control reactor (40X magnification) Control reactor (400X magnification)

Figure 6.8 Microscopic analysis of coupons stained with acridine orange (without pre-chlorination)

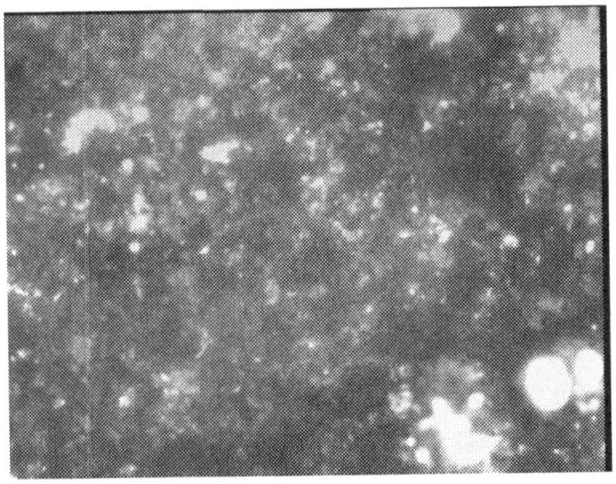

Control reactor

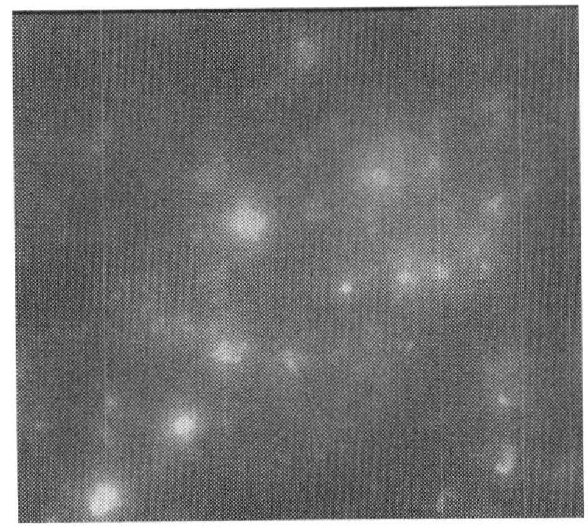

ClO_2 reactor

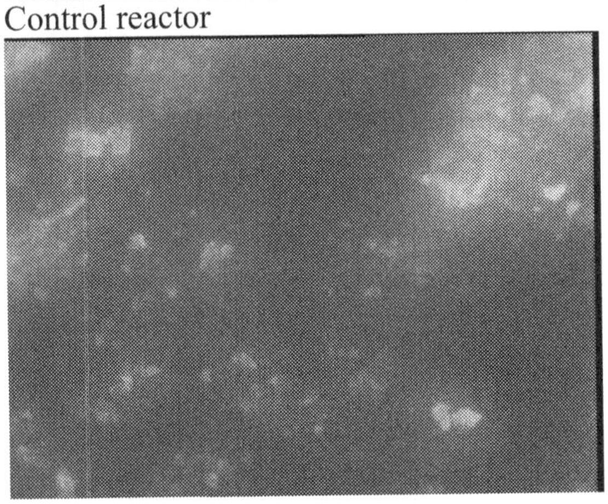

Cl_2 reactor

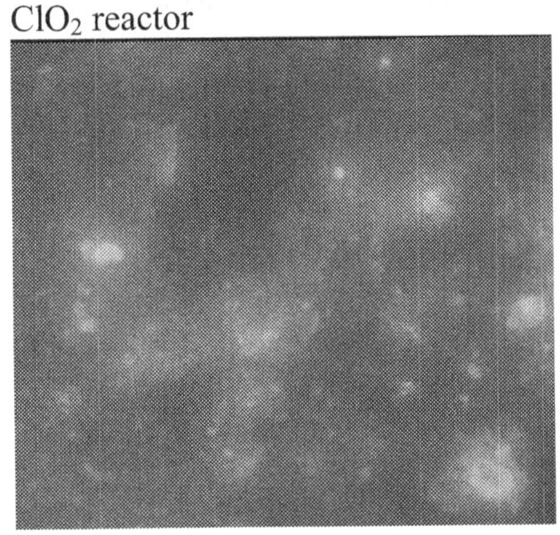

NH_2Cl reactor

Figure 6.9 Microscopic analysis of coupons stained with acridine orange (400X magnification) (without pre-chlorination)

Control reactor

ClO_2 reactor

Figure 6.10 Fouling of polycarbonate coupons caused by humic material

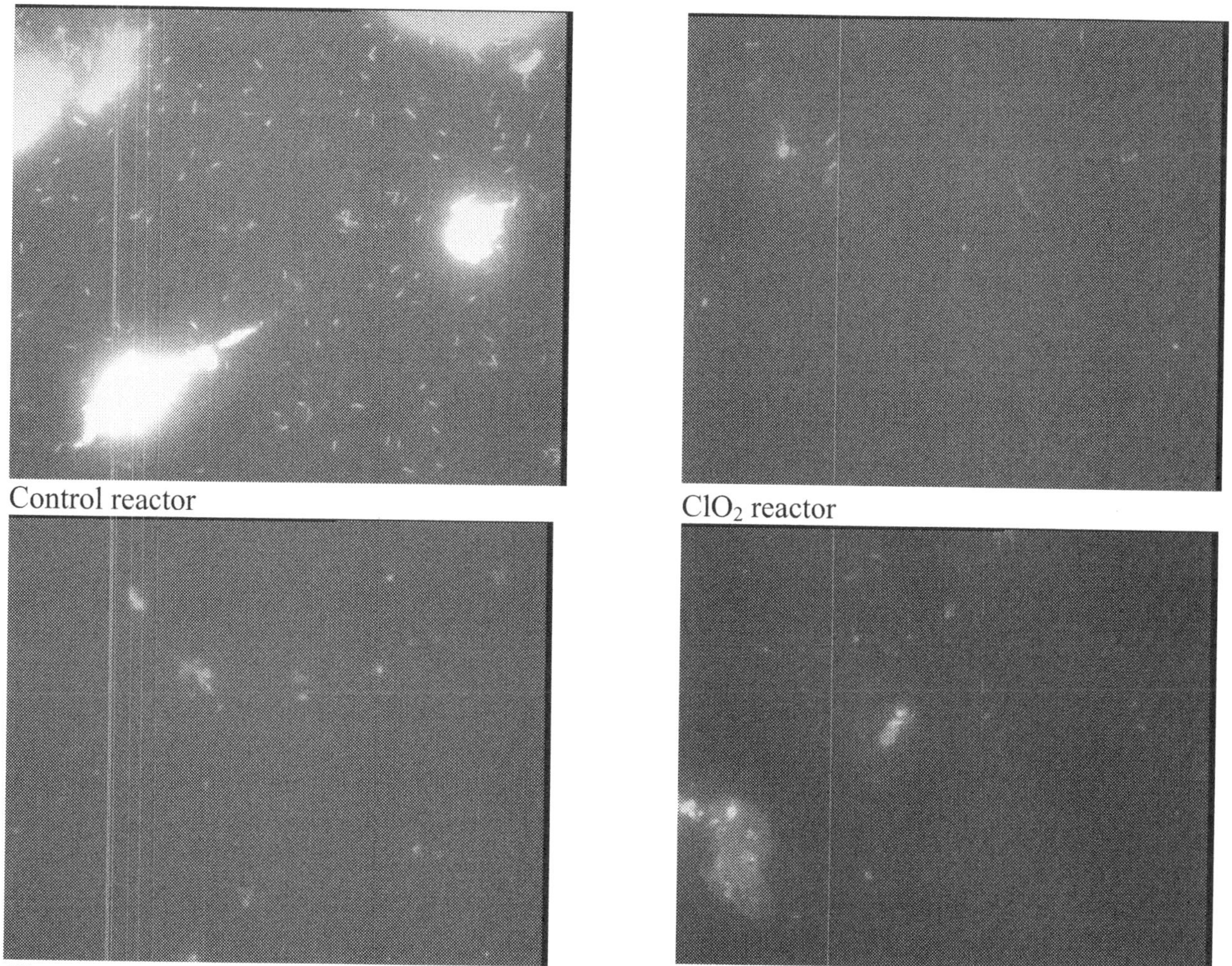

Figure 6.11 Microscopic analysis of coupons stained with acridine orange (400 X magnification) (with pre-chlorination)

Bacterial Identification

Overall, 107 bacteria from various samples were isolated and tested during phase 2. Both Gram negative (73 isolates or 68.2%) and Gram positive (34 isolates or 31.8%) bacteria were found (Table 6.13). Of the 107 isolates, 48 bacteria could be identified to at least the species level. Twenty-five isolates (23.3% of isolates) were identified as *Pseudomonas fluorescens*, whereas 11 isolates (10.3%) were identified as *Brevundimonas vescularis*, a pseudomonad-like bacterium (Table 6.13). Moreover, Gram positive cocci and aerobic Gram positive spore-forming rods, although not specifically identified to the species level, made up 16.8 (18/107) and 12.1% (13/107) of the isolates, respectively (Table 6.13). The common detection of spore-forming bacteria is not surprising because of their relative resistance to disinfection.

Table 6.13

Type of bacteria identified from Phase 2 and frequency of isolation

Bacterial Types or Species	Frequency of Isolation (percent)
Gram negative bacteria	68.2
Gram positive bacteria	31.8
Total	100.0
Pseudomonas fluorescens	23.3
Gram positive cocci	16.8
Gram positive spore forming rods	12.1
Brevundimonas vescularis	10.3
Gram positive rods (no spores)	2.8
Moraxella spp.	1.9
Agrobacterium radiobacter	1.9
Aeromonas salmonicida subs. *Achromogenes*	1.0
Other isolates	23.4
Total	100.0

[a] Only identifications with at least a 80% confidence level are reported in this table (based on the bioMérieux database).

Fewer bacterial types were found in raw water samples than in other samples (Table 6.14). The Gram negative isolates that appeared to be predominant in raw water samples could not be identified to the genus level. In addition, Gram positive bacteria were commonly found in raw water. A greater microbial diversity was observed in annular reactor effluent samples and in coupon samples, but there was no noticeable difference in bacterial diversity in these samples (Table 6.14). The most common bacteria in all samples with the exception of raw water samples were pseudomonads and pseudomonad-like bacteria (*Brevundimonas* was previously classified as *Pseudomonas*) (Table 6.14). Gram positive rods, and especially spore-forming ones, were found from each type of sample. Of note, one isolate of *Aeromonas salmonicida* was isolated from a disinfected (Cl_2) annular reactor effluent.

With the exception of raw water samples, in which very few bacterial types were found, no differences could be observed among the various samples with respect to the types of bacteria present. There were no apparent changes in population ratios with different disinfectants. The fact that fewer bacterial types were detected and identified in raw water samples, possibly because of the lack of sensitivity of the method used, suggests that bacterial regrowth in the annular reactor (and subsequent biofilm sloughing) was responsible for the increase in bacterial diversity detected in coupon and effluent samples.

Pseudomonads were by far the most common bacteria identified from the various samples. These bacteria are routinely detected in drinking water (Geldreich 1996). More specifically, *Pseudomonas fluorescens* is a ubiquitous bacterium that is commonly associated with soil and water. As its name implies, it produces a soluble, greenish fluorescent pigment. It is not considered to be pathogenic to humans, but can commonly form biofilms in distribution systems (Oliveira et al. 1994; Vieira and Melo 1995).

Brevundimonas, as previously mentioned, was originally classified as a *Pseudomonas*. It is an opportunistic pathogen, capable of causing septicemia. It is very common in soil, water,

Table 6.14

Locations of the isolation sites of the predominant heterotrophic bacteria isolated during Phase 2; isolation was done on R2A agar

Isolation Site	Predominant Bacterial Types or Species
Raw water (influent)	Gram negative rods (unidentifiable) Gram positive rods (non spore-formers) Gram positive rods (spore-formers)
Control Annular Reactor Effluents	*Pseudomonas fluorescens* *Agrobacterium radiobacter* *Moraxella* spp. Gram positive rods (non spore-formers) Gram positive rods (spore-formers)
Disinfected Annular Reactor Effluents	*Pseudomonas fluorescens* *Brevundimonas vescularis* *Aeromonas salmonicida* subs. *achromogenes* *Moraxella* spp. Gram positive cocci Gram positive rods (spore-formers)
Control Coupons	*Pseudomonas fluorescens* *Brevundimonas vescularis* Gram positive cocci Gram positive rods (spore-formers)
Disinfected Coupons	*Pseudomonas fluorescens* *Brevundimonas vescularis* Gram positive cocci Gram positive rods (non spore-formers) Gram positive rods (spore-formers)

and wasterwater. In one recent study, it was found to be one of the most dominant bacterial species identified from activated sludge (Juang and Morgan 2001).

The Gram negative genus *Aeromonas* is a fish and opportunistic human pathogen commonly occurring in surface waters supplying municipal drinking water systems (Kühn et al. 1997; Huys et al. 1997). It is increasingly recognized as an emerging waterborne pathogen, causing hemorrhagic septicemia or red sore disease and water-associated wound infections. As an opportunistic human pathogen, it has often been isolated from patients with diarrhea in the absence of other enteric pathogens, which suggests that it has some involvement in gastro-enteritis (Krovacek et al. 1992; Hanninen et al. 1995). *Aeromonas* can be found in biofilms in drinking water distribution systems, but its occurrence is rare in adequately disinfected systems (Chauret et al. 2001). The species *Aeromonas salmonicida* is better known as the causal agent of furunculosis in salmonid fishes, a debilitating and lethal disease encountered in aquaculture. Although *Aeromonas salmonicida* is not pathogenic to humans, it is closely related to many human and animal pathogens, and its presence suggests the possibility that other strains of aeromonads can be present.

Chemical Analysis

Total Organic Carbon

Average results for organic carbon analysis without and with pre-chlorination are shown in Table 6.15. Overall, the results show slight utilization of the humic material during acclimation and disinfection whether or not there was pre-chlorination. During experimentation without pre-chlorination there was a decrease in the amount of organic carbon in the reactor effluent from values observed in the UV clearwell (influent to the reactors). In addition there was no appreciable change in substrate utilization following secondary disinfection. Experimentation with pre-chlorination followed a similar trend. During this stage of testing there was also an observed increase in TOC during disinfection from acclimation values. This increase is attributed to an unexplained increase in BAC effluent organic concentration. During disinfection the organic loading in the effluent of the BAC filter column increased by an average 0.7mg/L from acclimation levels.

Disinfection By-Products

Average results for DBPs are shown in Table 6.16. There were no HAA data available for experimentation without pre-chlorination during the acclimation portion. During disinfection HAA results were the highest for the reactor that was disinfected with chlorine, followed by monochloramine and chlorine dioxide. This result is expected as chlorine dioxide has been shown to produce lower quantities of HAAs in full-scale distribution systems (Volk et al. 2002). Results from TTHM analysis resulted in a similar trend, with free chlorine producing the highest results. There were also no THMs measured in the influent water to the reactors.

During experimentation with pre-chlorination, results showed a significant production HAAs and THMs. During acclimation the 4 reactors exhibited similar levels of DBPs that were slightly above influent levels suggesting that there was additional formation within the model distribution system. TTHM analysis reports all THMs excluding chloroform. Analysis indicated a significant level of chloroform that is not typical. It was determined that samples were contaminated during transport.

Comparison with Phase 1

Figures 5.2 and 5.3 show suspended and biofilm HPC results for the reactors that received a synthetic cocktail of organic carbon. Figures 6.3 and 6.5 show suspended and biofilm HPC results for the reactors that received Aldrich humic material as their organic carbon source. In general, the figures show similar trends regardless of the carbon source.

Log reductions in HPC bacteria for each of the organic sources are illustrated in Figure 6.14. In the reactors receiving water supplemented with BOM, all three disintectants performed better at reducing suspended heterotrophic bacteria (Figure 6.14) when compared with the reactors receiving Aldrich humic material. Analysis of variance (ANOVA) tests ($\alpha=0.05$)

Table 6.15

Average TOC results

		No Chemical Disinfection	Chemical Disinfection
		TOC mg/L	TOC mg/L
A	Pre UV	2.63	2.41
	Post UV	2.53	2.46
	Control	2.50	2.22
	ClO_2	2.36	2.35
	Cl_2	2.36	2.17
	NH_2Cl	2.55	2.26
B	Pre UV	2.48	3.49
	Post UV	2.34	3.27
	Control	2.30	3.15
	ClO_2	2.22	3.18
	Cl_2	2.28	3.04
	NH_2Cl	2.24	3.31

A – No pre-chlorination.
B – Pre-chlorination.

Table 6.16

Average disinfection by-product results

		No Chemical Disinfection		Chemical Disinfection	
		HAA_6 µg/L	TTHM µg/L	HAA_6 µg/L	TTHM µg/L
A	Pre UV	-	0	6.3	0
	Post UV	-	0	8.7	0
	Control	-	0	0	0
	ClO_2	-	0	12.4	1
	Cl_2	-	0	34.7	4.5
	NH_2Cl	-	0	17.2	0
B	Pre UV	27.9	1.8	10.1	-
	Post UV	35.9	3.5	19.5	-
	Control	44.7	4.3	29.5	-
	ClO_2	43.0	4.1	32.7	-
	Cl_2	46.9	4.1	15.7	-
	NH_2Cl	46.6	4.1	20.2	-

A – No pre-chlorination.
B – Pre-chlorination.

Table 6.17

P-values from HPCs numbers that compare carbon source

	Suspended		Biofilm	
Reactor	Acclimation	Disinfection	Acclimation	Disinfection
Control	0.91	-	0.001	-
ClO_2	0.37	0.02	0.26	0.00015
Cl_2	0.59	0.00002	0.01	0.0094
NH_2Cl	0.30	0.0007	0.14	0.21

completed on pseudo steady state disinfectant values (Table 6.17) showed that the results were significantly different between organic carbon sources for each disinfectant investigated ($p \leq 0.02$). The suspended HPC results for the control reactor, during the entire duration of each experimental run, for each carbon source (BOM and Aldrich) were not shown to be significantly different ($p=0.91$).

An ANOVA tests completed on pseudo-steady state disinfectant values showed that the number of biofilm HPCs were significantly different between organic carbon sources for chlorine dioxide (ClO_2) and free chlorine (Cl_2) (Table 6.17). The number of biofilm HPCs for the control reactor, during the entire duration of each experimental run, for each carbon source (synthetic and Aldrich) were also shown to be significantly different ($p=0.001$). Biofilm bacteria were shown to be 1-log higher in the control reactor that received Aldrich humic material as the carbon source.

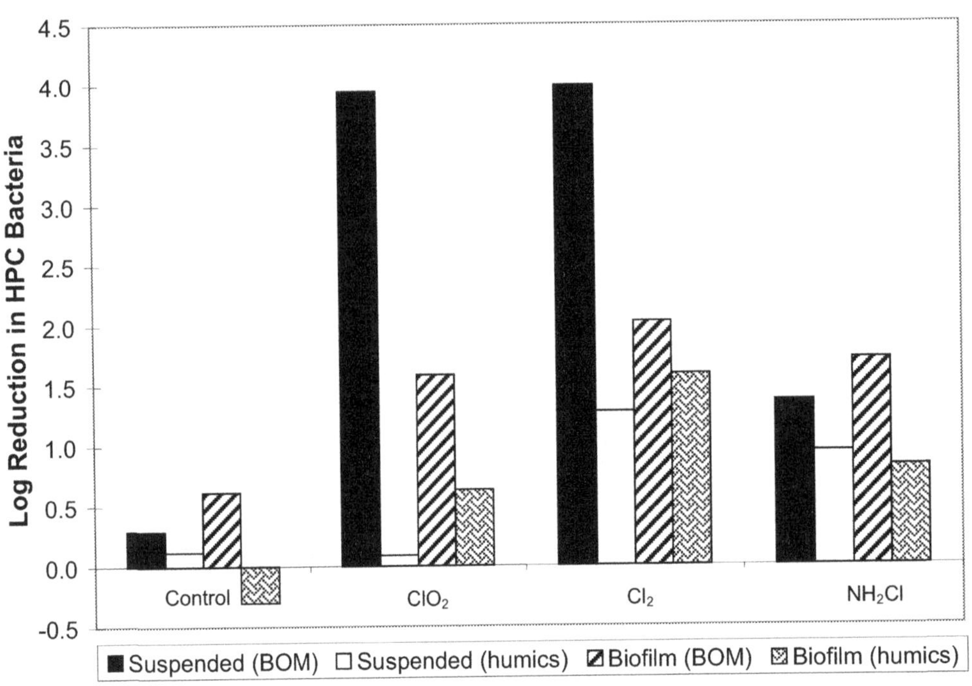

Figure 6.12 Log inactivation of HPCs for two organic sources

SUMMARY

This chapter evaluated the effect of UV and NOM coupled with pre-chlorination using a model distribution system. Results indicate that pre-chlorination coupled with UV can inhibit the development of biofilms and control the presence of bacteria in the bulk water. The combination of chlorine and UV disinfection was able to control the bacterial population in the UV clearwell. Pre-chlorination along with secondary disinfection with chlorine was able to completely eliminate suspended heterotrophic bacteria. All three disinfectants investigated were able to reduce suspended and biofilm HPCs below levels observed when pre-chlorination was not used. Chlorination prior to UV disinfection had a significant impact of both suspended and biofilm HPCs; they were the same after disinfection without pre-chlorination as they were during acclimation with pre-chlorination. The application of a secondary disinfectant after pre-chlorination was able to further reduce suspended and biofilm HPCs. The HPC results were evaluated to determine the detachment rate of the biofilm from the internal surface of the reactors as well as the disinfection coefficients for each of the disinfectants investigated. The detachment rates were calculated in the range if 0.001 to 0.01 h^{-1}, which were four times less than calculated in the previous phase of the research. Overall, these findings indicate that the use of chlorination prior to UV disinfection improves microbial control in distribution systems, and enhances the effectiveness of secondary disinfectants.

CHAPTER 7
POTENTIAL FORMATION OF CHEMICAL BY-PRODUCTS FROM UV DISINFECTION WITH CHLORINATION (PHASE 3)

INTRODUCTION

The objective of Phase 3 was to determine the impact of UV disinfection and pre-chlorination on natural surface water sources that have been filtered. A commercial source of organic carbon and three natural sources were treated with UV and a combination of UV and chlorine. Treated and control samples were collected and incubated at room temperature for a 25 day period. Samples were evaluated for TOC, DOC, DBPs, and organic by-products.

The fraction of dissolved organic carbon (DOC) that can be used as a source for growth by heterotrophic microorganisms is termed biodegradable dissolved organic carbon (BDOC) (Frías et al. 1992, Huck 1990, Block et al. 1992) and can be utilized by bacteria within days to months (Servais et al. 1989). BDOC is measured using bioassay techniques rather than chemical determination because organic matter is complex and difficult to detect at low concentrations (Frías et al. 1992, Servais et al. 1989). As a result various analytical methods are available for the determination of BDOC in drinking water using bioassay techniques.

Huck (1990) completed a survey of the various methods for measuring biodegradable organic (BOM) in drinking water and compares the methods that were available at the time of the review. Biodegradability measurement of aquatic NOM can be separated into biomass-based methods and DOC-based methods. The objective of both methods is to assess the removal of dissolved organic matter (DOM) by a treatment process or to predict the potential of water to support the growth of microorganisms (Huck 1990). Huck (1990) stated that the objectives do not imply the measurement of the same parameter and separates the objectives with two parameters and identifies the organic material that is producing the growth. Huck (1990) further divides the biomass-based methods into two groups based on the inoculum used; either one or more known organisms, or the indigenous bacteria from the water being tested. Biomass is then converted to an organic carbon concentration by means of a measured or assumed factor that is based on the method used. The DOC-based methods are based on the difference in DOC before and after incubation.

Oxidation of organic material has been shown to occur in the presence of UV light, as is the case in the natural environment. Organic matter extracted from a natural source was used in an UV disinfection study completed by Camper et al. (2001). As part of the study, oxidation of the humic material with ultraviolet light was investigated. The results indicated that the humic material was not significantly affected by UV disinfection with fluences in the upper range for water treatment (480 mJ/cm^2), approximately 10 times the fluence used in this current study. In the same UV disinfection study, the humic material when oxidized at very high fluences (25600 mJ/cm^2) experienced decreases in the hydrophobic fraction and aromaticity, and an increase of BDOC from 24 percent to 54 percent. As the fluence increased, the hydrophobic fraction and aromaticity both decreased and the BDOC increased.

Several studies have shown that ozone can increase the biodegradable organic matter, either quantification of BDOC (e.g., Graham 1999, Camel and Bermond 1998, Melin and Ødegaard 1999, Hozalski et al. 1999), or individual ozone by-products (e.g., Wienberg et al. 1993, Gagnon et al. 1997), and to some extent chlorine in a study conduced by Ellis et al. (2000), oxidation with UV light can increase the biodegradability of NOM. It is interesting to note that the BDOC was reported to increase as the fluence increased, although it required a very high fluence to observe significant changes (Ellis et al. 2000). With ozone it is known that NOM with a high percentage of low molecular weight material is not significantly oxidized (Hozalski et al. 1999), while the reverse is true for the high molecular weight material. Straight chain aliphatic aldehydes can also be formed as the partial oxidation products of the reaction between oxidants and NOM; the major ozonation by-products are aldehydes (Weinberg et al. 1993) and carboxylic acids (Gagnon et al. 1997). Once formed they can persist in the water and their concentrations may even increase following post disinfection. Fomaldehyde and bromate and other aldehydes are known to be of a health concern, but at low levels it is unlikely that they are a serious health hazard.

MATERIALS AND METHODS

Water Sources

Four water sources were chosen for this phase of the investigation. They represent three natural sources of organic matter and a synthetic source of organic matter. The natural source waters include post-filtered water from the Halifax Regional Water Commission (HRWC) and post-filtered water from the Region of Ottawa-Carleton, and post settled water from East Bay Municipal Utilities District (EBMUD). The initial TOC concentrations of these waters were 2.1, 3.4, and 3.6 mg/L, respectively. Commercial humic acid (Aldrich Chemical Company) is the synthetic organic source chosen. These humic acids are extracted from peaty soils and are known to be different from natural aquatic NOM sources (Goel et al. 1995).

Source Water Preparation

A concentrated stock (210 mg/L) commercial humic acid was diluted with tap water passed through GAC and BAC filters to a concentration of 2.5 mgL. The natural sources were used directly from the water collected.

Batch Biodegradation Studies

BDOC determination was competed using a method that had been modified from one described by Goel et al. (1995). This method was chosen in order to frequently monitor changes in DOC over time. Each source was treated with UV (40 mJ/cm^2) and a combination of UV and chlorine (1 mg/L) as indicated in Figure 7.1 and transferred into eight flasks, which had been heated at 250°C for 8 hours, for BDOC testing. Initial HPCs, TOC, DOC, THMs, HAAs, aldehydes and carboxylic acids were determined. Subsequent analysis took place on day 1, 2, 4, 6, 10, 15, and 25 of the study. Bulk water samples for BDOC analysis were collected in 300-mL glass bottles (Corning Inc., Acton, MA) containing 1% (w/v) sterile sodium thiosulfate. Water was irradiated with the UV disinfection system that was described in detail in Chapter 3.

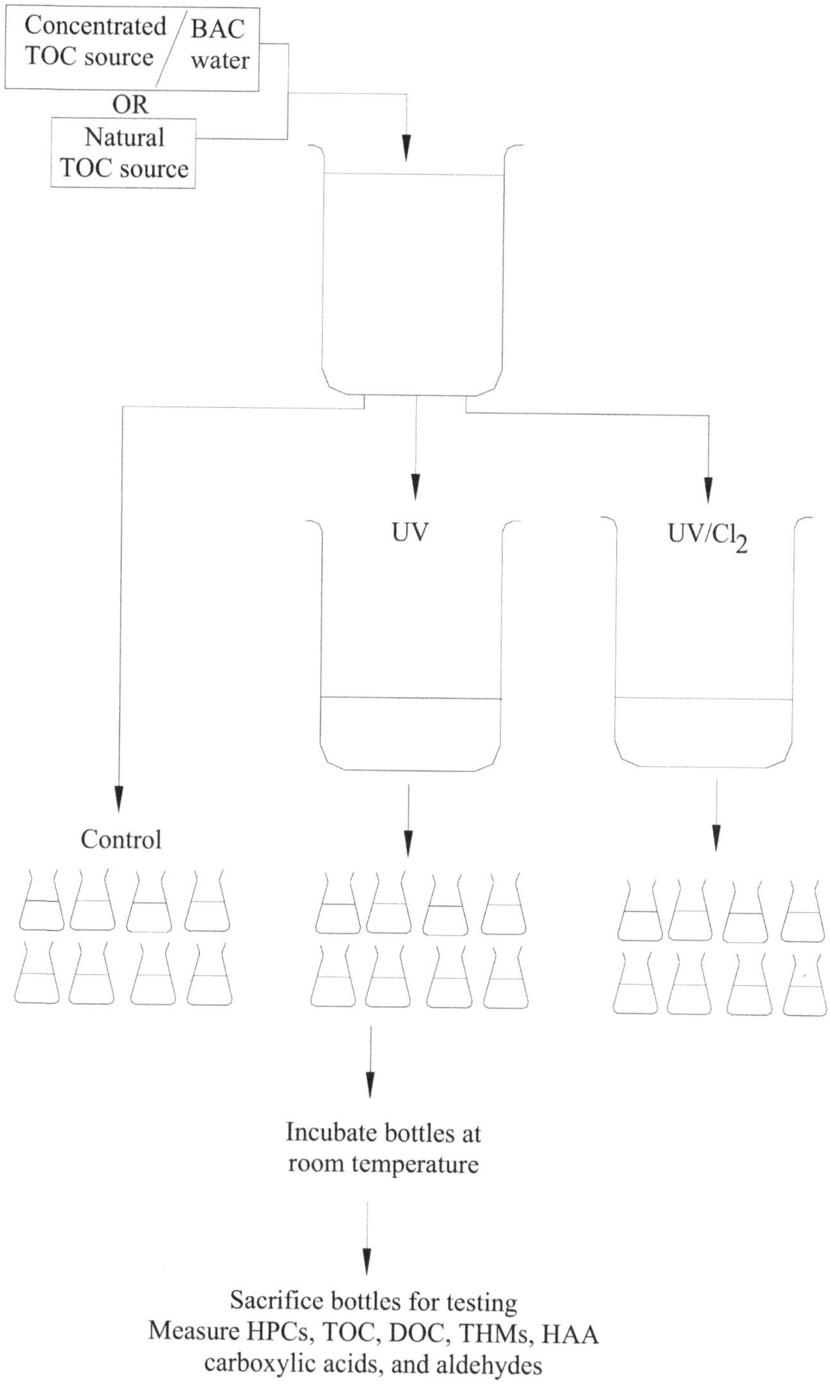

Figure 7.1 Batch experiment set up for Phase 3 biodegradation study

RESULTS AND DISCUSSION

Dissolved Organic Carbon and Microbial Growth

Analysis completed for the control experiments resulted in a slight decrease in DOC from initial values measured at day zero for the natural waters as well as the synthetic water (Figure 7.2a). Some degradation was expected to occur, as the amount of organic carbon available for microbial growth is constant while there was an increased amount of microbial cells utilizing the carbon source (Figure 7.3a). Overall, the DOC for the control experiments remained virtually the same, as the calculated BDOC values were small (0.13-0.80 mg/L) over the experimental trial (Table 7.1). BDOC was calculated as the difference from the initial (t=0) to the final (t=25) DOC measured. Microbial growth varied significantly depending on the source and initial bacterial concentrations (Figure 7.3a). Heterotrophic bacteria reached a maximum concentration on the order of 1×10^5 by day 10 of the study after which pseudo steady state was achieved. After day 15 there was also no observed decrease in DOC. The absence of additional heterotrophic bacteria growth after day 10 may indicate the absence of any additional available carbon.

Following oxidation with UV light there was a significant increase in DOC concentration for post-filtered water collected from Halifax of approximately 6 fold. Water collected from Ottawa and EBMUD experienced only slight increases in DOC from control levels. It was suspected that analytical problems caused the the increased concnetration in DOC after oxidation with UV, however this did not occur on standard solutions that were quantified over that period. Although there was no appreciable reduction in DOC during the initial incubation days, there was an observed degradation of DOC over the entire experimental trial for the natural and synthetic waters (Figure 7.2b). The waters oxidized with UV light experienced a general increase in biodegradability as indicated from BDOC calculations (Table 7.1). However, the results are slightly biased, specifically in the case of Halifax water, due to the increase in initial DOC from the control results. Microbial growth followed similar trends regardless of the source and initial bacterial concentrations (Figure 7.3b). Heterotrophic bacteria reached a maximum concentration on the order of 1×10^5 by day 5 of the study after which pseudo steady state was achieved.

Water treated with a combination of UV and chlorine showed similar trends (Figure 7.2c) as the waters that were treated with UV alone; there was a general increase in DOC for the natural waters from initial levels measured in the control experiments. Microbial growth followed similar trends regardless of the source and initial bacterial concentrations (Figure 7.3c). Heterotrophic bacteria reached a maximum concentration on the order of 1×10^5 by day 5 of the study after which pseudo steady state was achieved.

Table 7.1

BDOC results from batch biodegradation studies

	Treatment		
Water	Control	UV	UV/Cl_2
Halifax	0.128	9.432*	2.613*
Ottawa	0.516	1.362	2.943
EBMUD	0.808	0.439	1.824
Aldrich	-0.308	0.520	0.496

* This experimental trial experienced a significant increase in DOC (from initial levels).

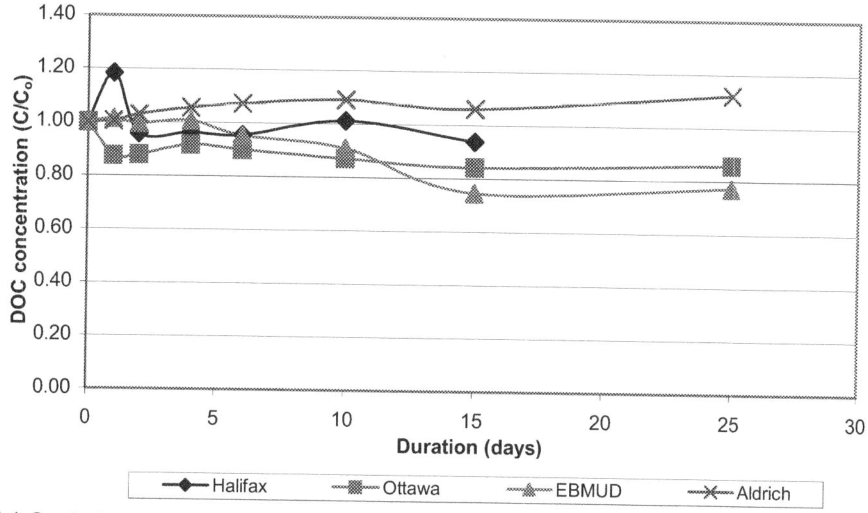

(a) Control

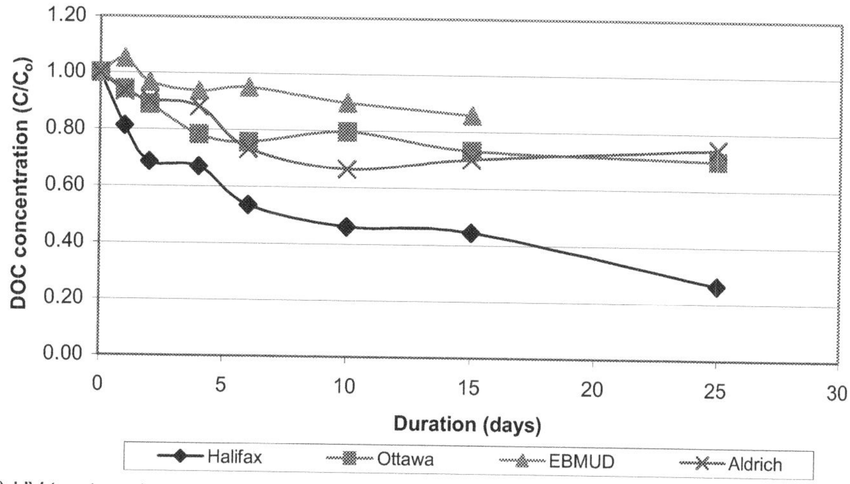

(b) UV treatment

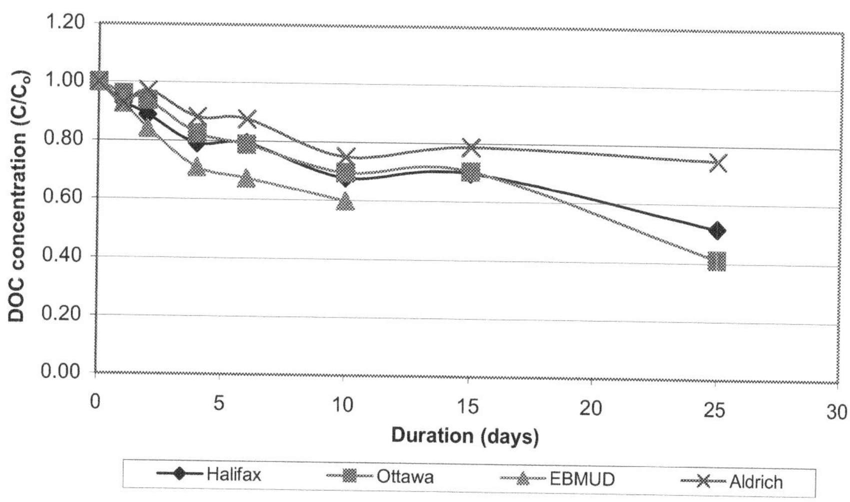

(c) $UVCl_2$ treatment

Figure 7.2 DOC results from batch biodegradation studies

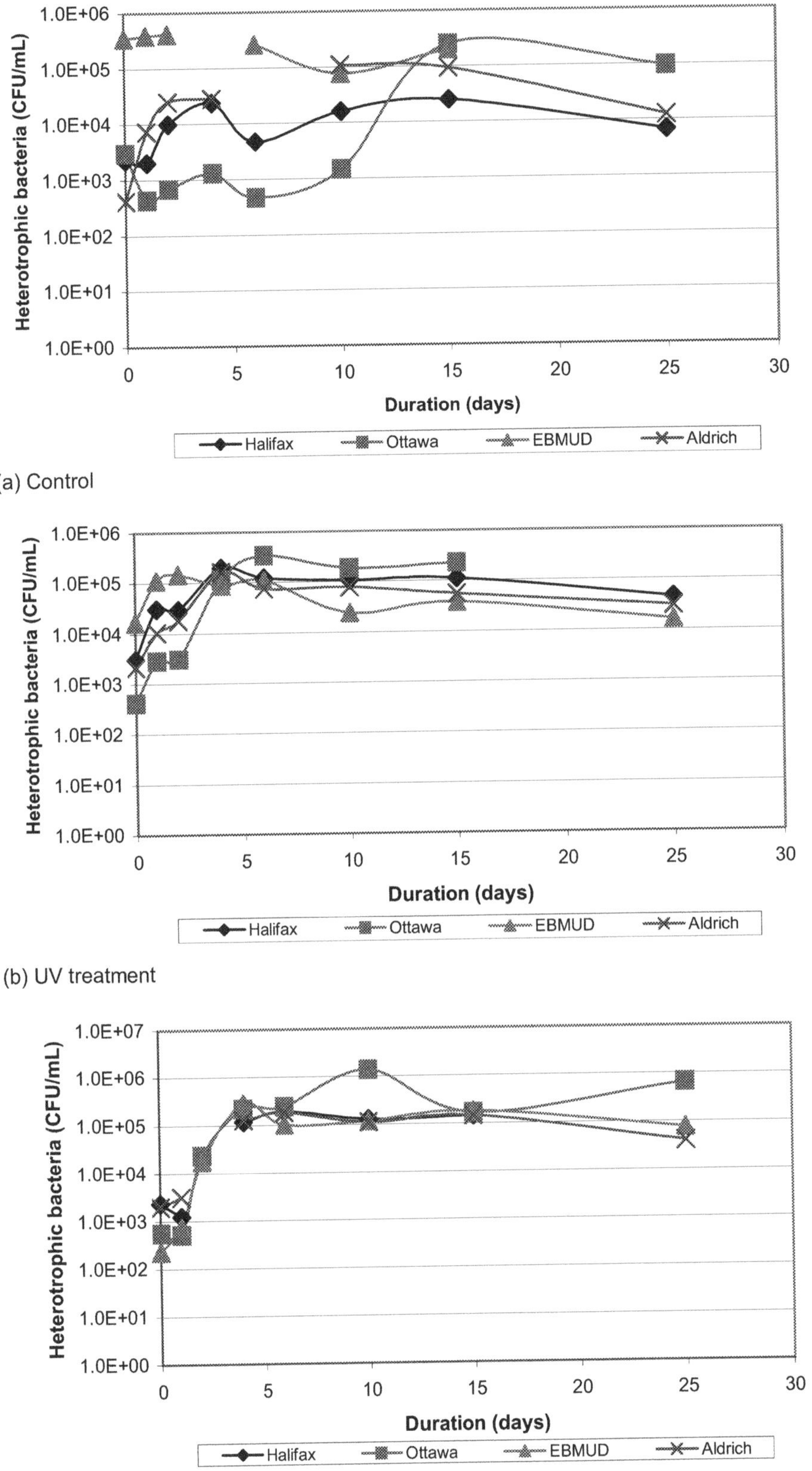

(a) Control

(b) UV treatment

(c) $UVCl_2$ treatment

Figure 7.3 Microbial growth results from batch biodegradation studies

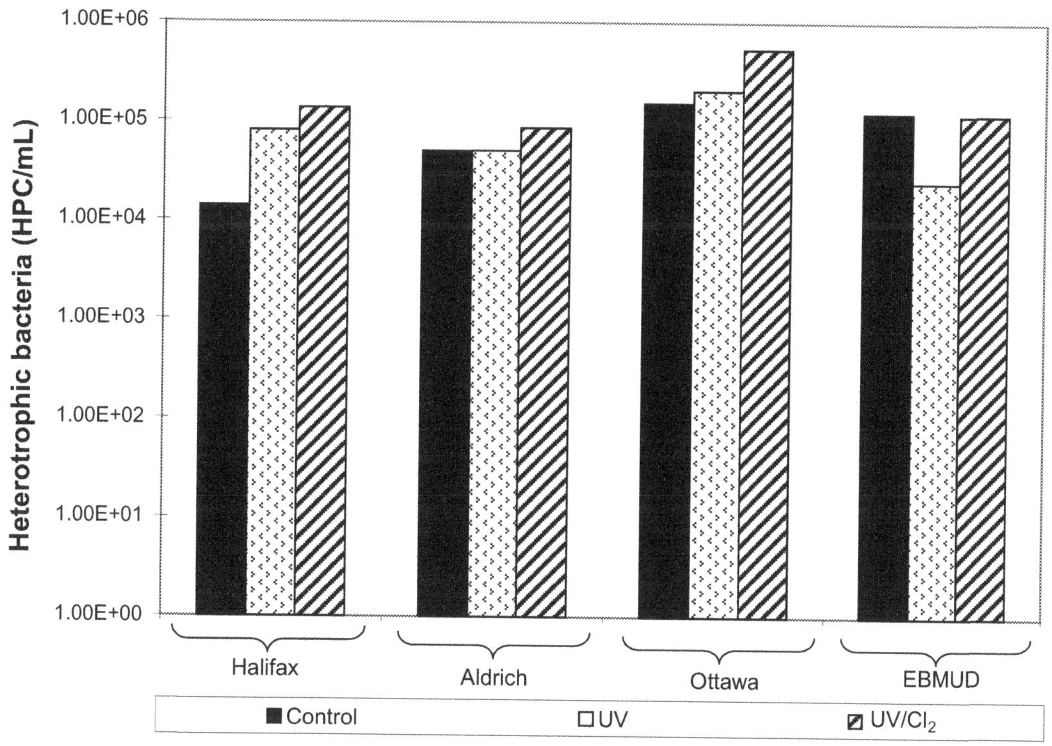

Figure 7.4 Mean number of heterotrophic bacteria during days 10 to 25

Camper et al. (2001) reported an increase in BDOC from 24-54% after UV oxidation although with a fluence of approximately 25000 mJ/cm^2. The results presented in Figure 7.4 appear to contradict the microbial counts for Phase 2; in that during Phase 2 heterotrophic counts decreased following UV/Cl_2 treatment. However during Phase 3 the water was immediately quenched with thiosulfate, such that the effect of oxidation could be examined rather than the biocidal effects of chlorine, which were examined in Phase 2. In practice, these biocidal effects would have a significant effect on growth.

Carboxylic Acids and Aldehydes

Analysis was completed on the initial day (t=0) and final day (t=25) of experimentation for the production of carboxylic acids and aldehydes after oxidation with UV and a combination of UV and chlorine. Results from carboxylic acid analysis showed the production of acetate, pyruvate, formate, glycolate, and butyrate in samples analyzed on day 0. The same compounds were found on day 25 along with ketone, which was found in two samples. These results proved to be very inconsistent between duplicate samples for half of the samples analyzed and there were no trends observed with respect to formation and decomposition from day 0 to day 25.

Results from aldehyde analysis are illustrated in Figures 7.5 to 7.8. All four waters investigated showed and increased production of various aldehydes as the result of increased oxidation from the control experiments, the most prevalent being acetaldehyde and formaldehyde, which is consistent with ozone work done by Weinberg et al. (1993). Post-filtered water from Halifax showed the highest production of aldehydes with increased oxidation with the maximum achieved after oxidation with UV/Cl_2 (Figure 7.5). However, this was also the case in the BDOC analysis of Halifax water where there was a significant increase in TOC following oxidation.

An increase in the concentration of acetaldehyde, of all four waters, from control levels was observed following UV oxidation, and an additional increase was observed following oxidation with UV/Cl_2. Additional aldehydes have been shown to form as a result of chemical oxidants (Weinberg et al. 1993). Oxidation with UV alone did not effect the production of formaldehyde, as it did not change appreciably from control levels. However, oxidation with the combination of UV and chlorine resulted in noticeable increases in the concentration of formaldehyde. In addition propionaldehyde and butyraldehyde were only detected after oxidation with the combination of UV and chlorine. This increase of formaldehyde and the formation of propionaldehyde and butyraldehyde could possibly be controlled by the chlorine alone, the combined oxidation of UV and chlorine, or as a result of the interactions between UV and chlorine. A small amount of benzaldehyde was also detected in the UV/Cl_2 sample of Ottawa water.

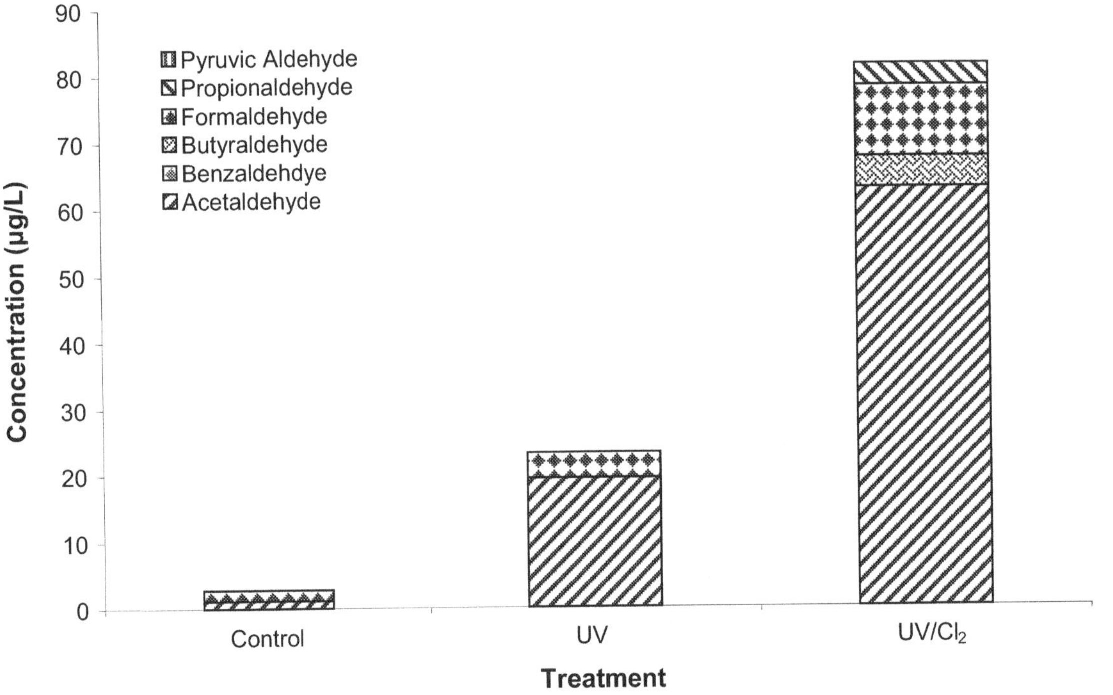

Figure 7.5 Aldehyde formation as a function of oxidation type for post-filtered water from Halifax

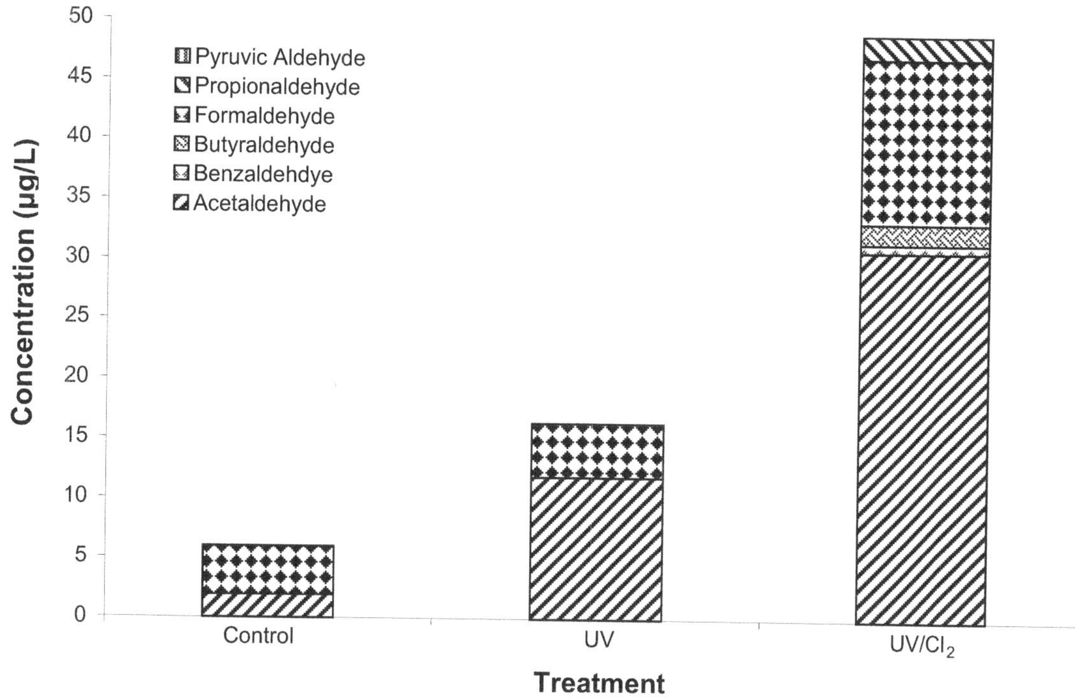

Figure 7.6 Aldehyde formation as a function of oxidation type for post-filtered water from Ottawa

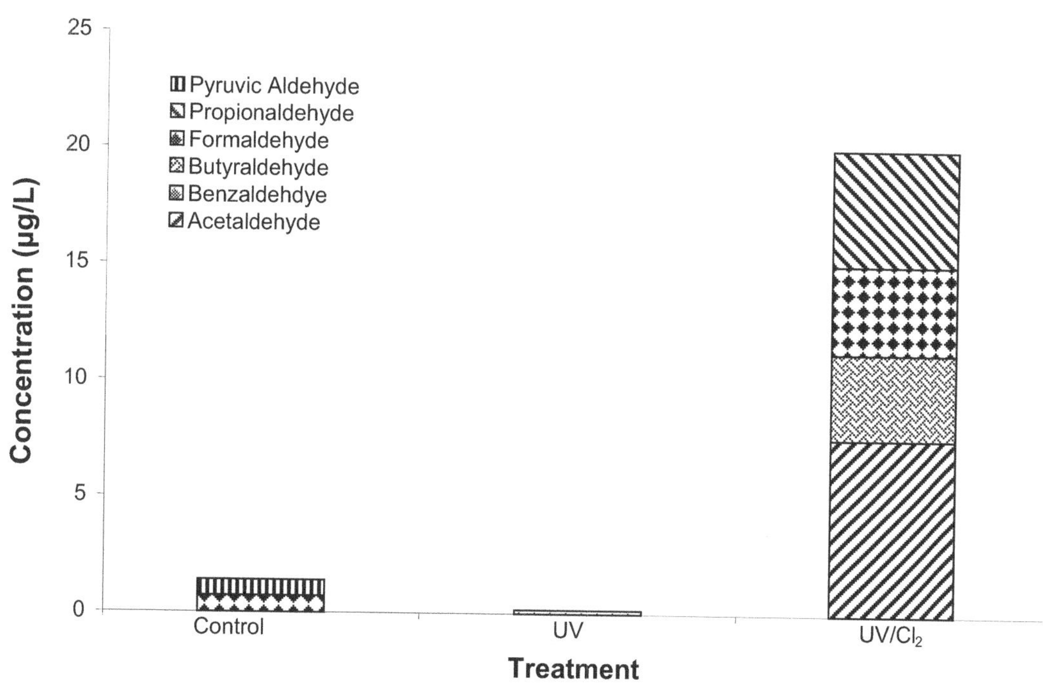

Figure 7.7 Aldehyde formation as a function of oxidation type for post-settled water from EBMUD

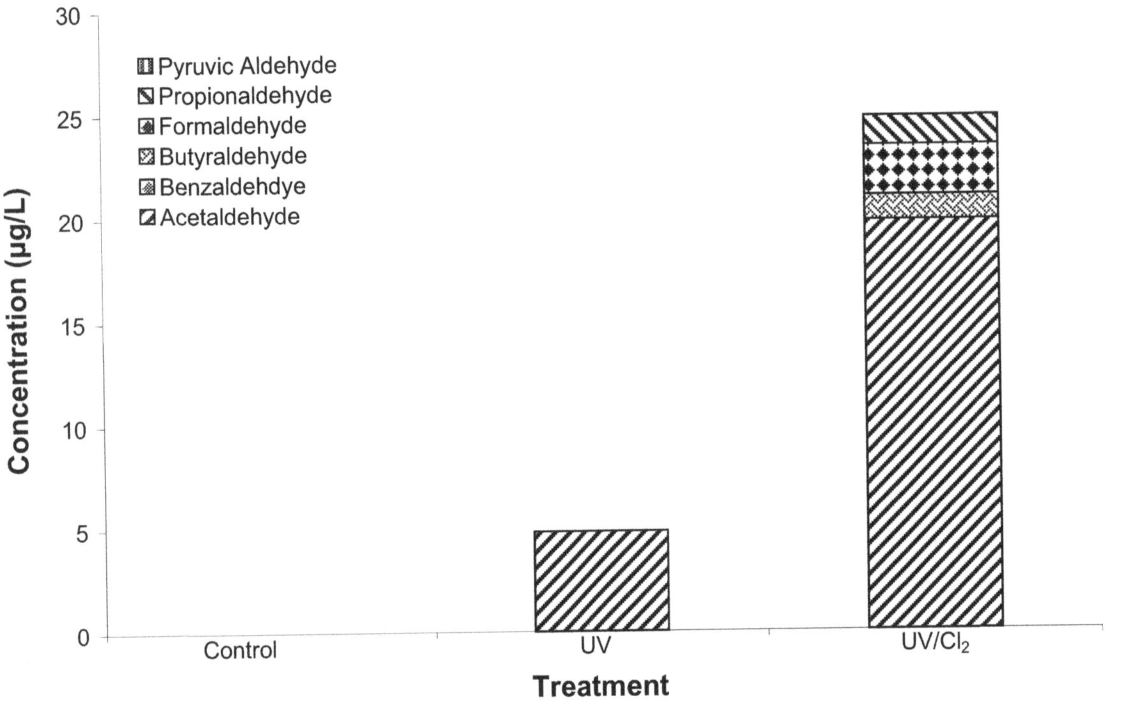

Figure 7.8 Aldehyde formation as a function of oxidation type for Aldrich humic acid

Disinfection By-Products

Analysis was completed on the initial day (t=0) and final day (t=25) of experimentation of the production of THMs and HAAs after oxidation with UV and a combination of UV and chlorine. Results for HAA analysis are reported in Table 7.2. THM results are reported in Table 7.3, excluding results for chloroform as the samples were contaminated with chloroform during sample storage. HAA_6 experienced an observed increase as a result of oxidation with UV light and additional increase resulting from UV/Cl_2 oxidation (Figure 7.9) for the three natural waters. Aldrich humic acid did not experience an increase in HAA production as a result of UV oxidation alone. Over the entire incubation time of 25 days there was also an observed decrease in HAA as a result of biological degradation, which is consistent with results reported by others (Chen and Weisel 1998, Thomas et al. 2000). HAA results are below the stage 1 and stage 2 maximum contaminant level (MCL) of 60 µg/L and 30 µg/L, respectively, set by the USEPA (AWWA and ASCE 1998).

Table 7.2

HAA_6 results from batch biodegradation studies (μg/L)

	Control		UV		UV/Cl_2	
	Day 0	Day 25	Day 0	Day 25	Day 0	Day 25
Halifax	0.81	0.11	3.49	1.06	7.19	11.3
Ottawa	2.33	1.37	4.71	4.58	28.0	1.78
EBMUD	0	0	0.25	0	5.95	3.03
Aldrich	0	0	0	0	0.96	0.64

Table 7.3

TTHM results from batch biodegradation studies (μg/L)

	Control		UV		UV/Cl_2	
	Day 0	Day 25	Day 0	Day 25	Day 0	Day 25
Halifax	0	2	3	2	3	2
Ottawa	6	3	6	3	6	3
EBMUD	0	0	0	-	3	-
Aldrich	0	0	0	0	1	0

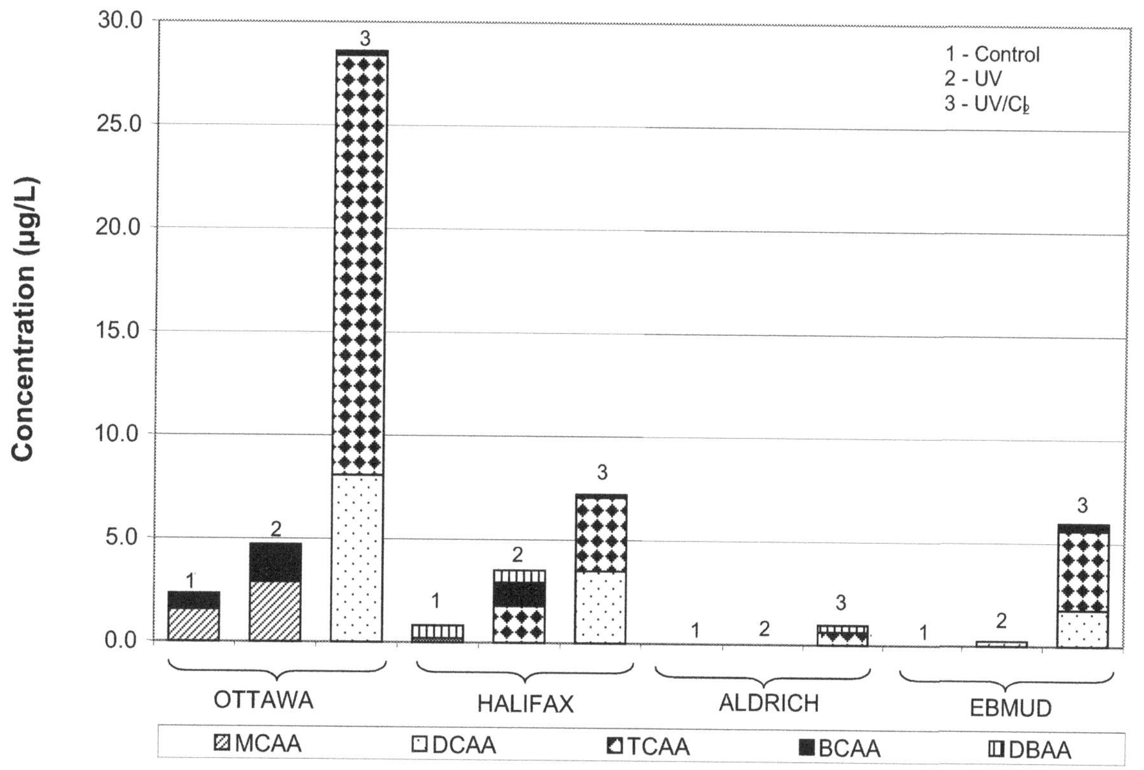

Figure 7.9 HAA formation as function of oxidation type

SUMMARY

Phase 3 investigated the impact that oxidation of NOM has on biological growth and biodegradability. Overall, there was an increased potential for biodegradation with increasing oxidation for the natural water sources, with an increased BDOC from control levels. The commercial organic source experienced similar biodegradation capacity for both treatments and was greater than the control. Growth rates were also affected with increasing oxidation. Although HPC levels after UV and UV/Cl_2 treatment were below levels observed in control experiments, maximum growth was achieved after day five. Maximum growth was achieved after day 10 in the control experiments. This may indicate that the organic material present is more available for bacterial growth. There was also an increased production of aldehydes after oxidation for all four waters investigated. Acetaldehyde increased after oxidation with UV alone and a combination of UV and chlorine. Oxidation with a combination of UV and chlorine resulted in the production of propionaldehyde and butyraldhyde, along with an increase in the concentration of formaldehyde.

CHAPTER 8
IMPACT OF FOULING BY HUMIC MATERIAL ON THE EFFICACY OF UV DISINFECTION

INTRODUCTION

The presence of organic material and particulate matter in source water play an important role in the disinfection efficiency of UV light. The presence of humic material in source water that is irradiated can be a limiting factor on the biocidal effect of UV light. Qualls et al. (1983) reports that organic particles can become a major limiting factor in disinfection. This limitation is attributed to humic substances being the most highly UV absorbing compounds in natural waters (Lund and Hongve 1994). Suspended particles in water can play two roles, they can absorb and scatter UV light and they can harbour bacteria to protect them from light (Qualls et al. 1983). Fouling of quartz sleeves has also been shown to be a major limiting factor in the use of low pressure UV light and has been attributed to a combination of hydraulics, heat, and water quality effects (Sheriff and Gehr 2001).

The primary objective of this chapter was to determine the ability of a medium pressure UV lamp, operated at 45 mJ/cm^2, to inactivate MS 2 after a quartz tube has been fouled with organic rich source water, with and without pre-chlorination. To this end the inactivation of MS 2 was determined in the fouled and clean tube, in the presence and absence of organic rich water. An additional objective was to assess the effect of lamp age on the ability of UV to inactivate MS 2.

MATERIALS AND METHODS

Specific methods for the operation of the UV disinfection assembly are described in Chapter 3. Methods for MS 2 evaluation are described in Chapter 4. In contrast to Chapter 4, however, calibration the of lamp took place following exposure to an organic loading of 2.5 mg/L for a period of 12 weeks, with and without pre-chlorination (i.e., after an experimental trial had been completed in Phase 2). At the end of these trials MS 2 inactivation studies were completed in a manner similar to the procedure described in Chapter 4.

RESULTS AND DISCUSSION

Based on initial experimentation discussed in Chapter 4 a fluence range of 10 – 45 mJ/cm^2 represents the minimum and maximum that can be achieved with the UV disinfection assembly under the arrangement and conditions that were investigated. A fluence of 40 mJ/cm^2 represents a 2-log inactivation of MS 2 bacteriophage (Blatchley et al. 2000) and is considered to be in the range for drinking water treatment applications (Linden and Mofidi 1999). UV fluence ranges from 30 to 80 mJ/cm^2, which represents a low fluence for bacterial inactivation and high fluence for viral disinfection (Linden and Mofidi 1999).

The UV disinfection system had been previously calibrated and fluence was determined. Results showed that after dispersion and kinetics were accounted for within the UV reactor, the modeled results for a photolytic reaction length of 0.4 cm predicted a fluence that was in general agreement with NWRI design curves for inactivation MS 2 bacteriophage. MS 2 inactivation

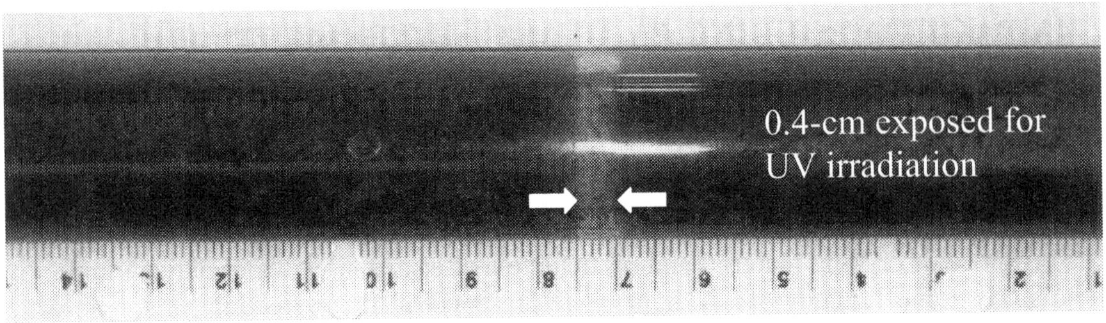

Figure 8.1 Organic fouling of quartz tube (without pre-chlorination)

trials conducted with a lamp that had been continuously operated for a 5-month period achieved an average inactivation 2-log at 45 mJ/cm^2.

Subsequent to calibration the lamp was allowed to operate at 45 mJ/cm^2 with an organic loading of 2.5 mg/L for a period of 12 weeks, with and without pre-chlorination, after which MS 2 inactivation studies were completed. Figure 8.1 shows the extent of organic fouling in the quartz tube that occurred as a result of the organic loading without pre-chlorination. The quartz tube, with pre-chlorination, was considerably less fouled on the portion that was covered, the uncovered 0.4-cm was similar to that shown in Figure 8.1.

The presence of humic material in the source water plays a significant role in the efficiency of UV light for inactivation. MS 2 inactivation trials conducted with the fouled quartz tube (without pre-chlorination) in place achieved an average inactivation 1.46-log at 45 mJ/cm^2 (Table 8.1a). Trials conducted with the fouled tube in place and spiked with 2.5 mg TOC/L of humic material achieved an average inactivation of 1.3-log. Similar studies were completed with a clean tube in place, which resulted in an average MS 2 inactivation of 3-log without the addition of humic material and 1.6-log with humics spiked at 2.5 mg/L. Experiments conducted with the clean tube resulted in MS 2 inactivation 1.5-log higher than the trials conducted with humic material present in the source water. In trials conducted with the fouled tube MS 2 inactivation was slightly higher (0.2-log) than when humic material was present in the source water (Figure 8.2a).

MS 2 inactivation trials conducted with the fouled quartz tube (with pre-chlorination) in place achieved an average inactivation 0.71-log at 45 mJ/cm^2 (Table 8.1b). Trials conducted with the fouled tube in place and spiked with 2.5 mg TOC/L of humic material achieved an average inactivation of 0.25-log. Similar studies were completed with a clean tube in place, which resulted in an average MS 2 inactivation of 1.93-log without the addition of humic material and 0.78-log with humics spiked at 2.5 mg/L. Experiments conducted with the clean tube resulted in MS 2 an inactivation of 1.2-logs higher then the trials conducted with humic material present in the source water. In trials conducted with the fouled tube MS 2 inactivation was slightly higher (0.46-log) than when humic material was present in the source water (Figure 8.2a).

Table 8.1a

Log inactivation of MS 2 under clean and fouled tube conditions (no pre-chlorination)

Fluence (mJ/cm^2)	Fouled tube	Fouled tube with humics	Clean tube	Clean tube with humics
10.02	-1.18	-0.63	-1.15	-0.25
16.82	-1.35	-0.96	-1.73	-0.29
24.95	-0.90	-1.41	-1.82	-0.89
33.90	-1.20	-1.48	-1.23	-1.42
45.84	-1.46	-1.30	-3.10	-1.49

Results presented in Figure 8.2a shows a maximum clean tube inactivation of about 3-logs. This result is 1-log over that experienced in similar MS 2 inactivation experiments conducted. The increased MS 2 inactivation in this one experiment may be the result of the number of influent MS 2 being about 0.8-1.2 logs higher than the other experiments. If the influent number of MS 2 present was similar to other experiments the overall results may have been similar to those illustrated in Figure 8.2b.

CONCLUSIONS

Overall, the results suggest that short-term organic fouling of a quartz tube has a significant impact on the disinfection capacity of a medium pressure UV lamp when humic material is not present in the source water. The presence of organics in the source water also plays a significant role in reducing the capacity of UV for bacterial and viral disinfection when experiments were completed with the clean tube in place. With the fouled tube in place the disinfection capacity is decreased slightly. At this time lamp age does not seem to have any effect on the efficiency of UV disinfection. Also, the use of pre-chlorination does not seem to have any effect on fouling as inactivation results were similar to MS 2 inactivation when pre-chlorination was not used.

Table 8.1b

Log inactivation of MS 2 under clean and fouled tube conditions (pre-chlorination)

Fluence (mJ/cm^2)	Fouled tube	Fouled tube with humics	Clean tube	Clean tube with humics
10.02	-0.01	0.25	0.25	-0.13
16.82	0.31	0.12	0.20	0.01
24.95	-0.23	0.21	-0.74	-0.15
33.90	0.00	0.11	-0.90	-0.10
45.84	-0.71	-0.25	-1.93	-0.78

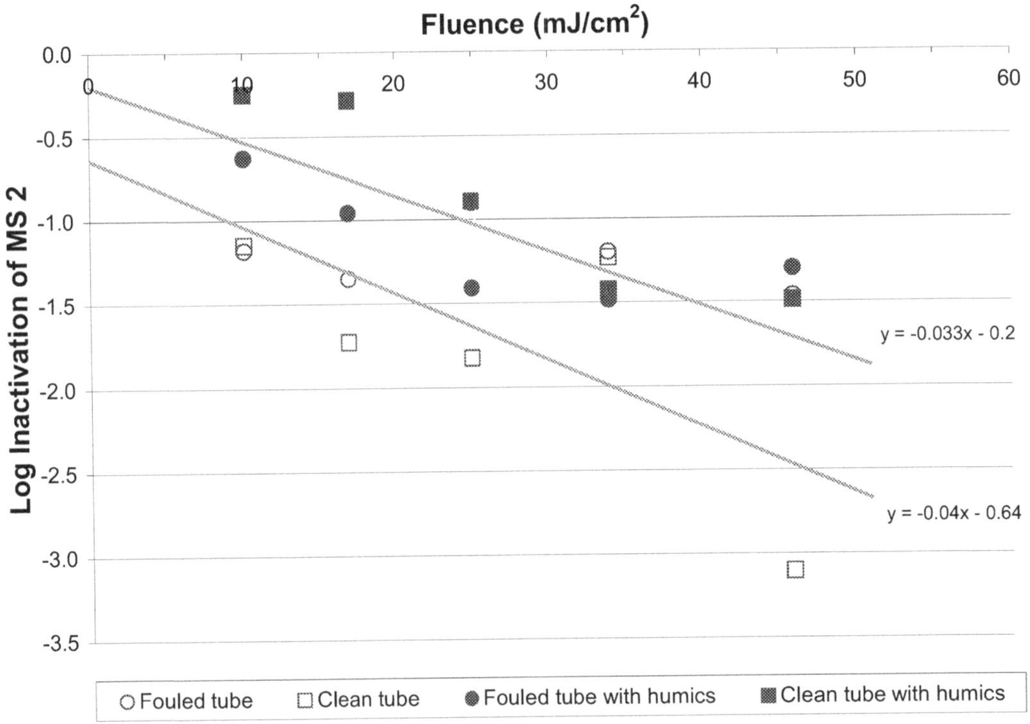

Figure 8.2a MS 2 inactivation under fouled and cleaned conditions (no pre-chlorination)

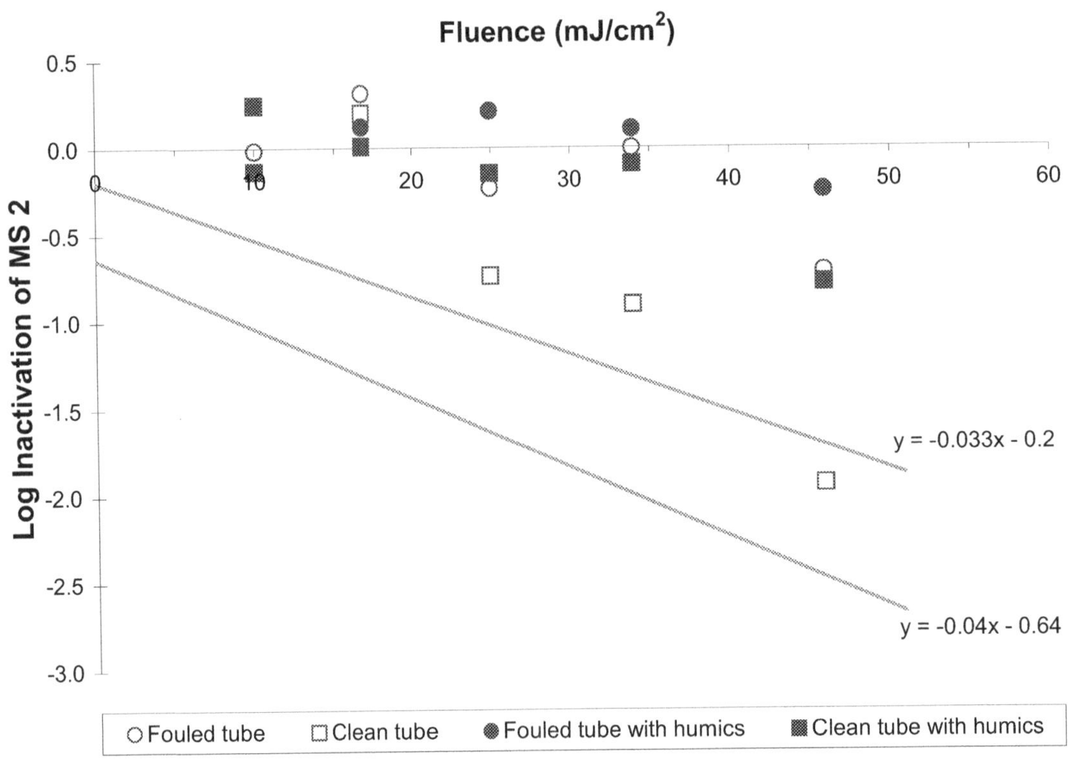

Figure 8.2b MS 2 inactivation under fouled and cleaned conditions (pre-chlorination)

CHAPTER 9
PROJECT SUMMARY AND RECOMMENDATIONS

PROJECT SUMMARY

This investigation evaluated the impact of UV disinfection on biofilm accumulation in drinking water distribution systems. To address this broad topic area the research was conducted in three phases. Phase 1 evaluated the synergistic effects of UV coupled with secondary disinfectants using a model distribution system. The secondary disinfectants studied in Phase 1 and throughout the project were free chlorine, chlorine dioxide and monochloramine. Phase 2 examined the biodegradability of humic material that has been treated with UV and chlorine. Phases 1 and 2 were conducted at bench-scale using annular reactors (ARs) to monitor bacterial regrowth. Phase 3 characterized the potential for biological regrowth by quantifying the biodegradable dissolved organic carbon and organic by-products produced following UV/Cl_2 catalysis of post-filtered water obtained from three water utilities. From these three phases several key issues can be drawn from the research, namely: the impact UV disinfection has on biological stability, the impact of chemical disinfectant following UV disinfection, and the role that NOM has on the effectiveness of UV disinfection and biological stability.

Impact of Ultraviolet Disinfection

Disinfection with UV light has been shown to enhance the potential for microbial control in drinking water distribution systems. The use of UV prior to the application of a secondary disinfectant resulted in enhanced microbial inactivation in the bulk water phase. However, the established biofilm maintained its ecological properties and resulted in similar behavior to chemical disinfection. At the high chemical disinfectant level, UV treatment resulted in significantly lower suspended heterotrophic bacteria for the three chemical disinfectants. Free chlorine resulted in significantly lower numbers of bulk heterotrophic bacteria at the low disinfectant concentration. These results suggest that there was synergy between primary UV disinfection and secondary chemical disinfection, but that the potential synergy depends on the chemical disinfectant concentration. For low secondary disinfectant residuals, the results indicate that a synergistic interaction exists between UV and the secondary disinfectants. This synergic interaction was not observed, or at least masked, for the high disinfectant residual concentrations. This could indicate that for higher concentrations of chemical disinfectants UV disinfection plays a minimized role in the control of suspended bacterial populations. When used without a secondary disinfectant UV alone was unable to control the proliferation of both suspended and biofilm bacteria. Practically, these results would indicate that a dual-barrier disinfection system should be considered for pathogen protection at the plant and distribution system.

Impact of Secondary Disinfection

Chlorine dioxide performed better than free chlorine on an equivalent Ct basis for the inactivation of suspended bacteria, whether or not UV disinfection was employed. Biofilm bacteria reacted similarly regardless of the disinfectant type or dose. Chlorine dioxide and free chlorine performed better at reducing suspended bacteria when UV disinfection was used prior to secondary disinfection. UV disinfection appears to impact the suspended bacteria in such a way that they become more vulnerable to chemical disinfection. With an increase in the secondary disinfectant dose there was a corresponding increase in suspended bacteria reductions, although these reductions were not found to be significantly different among chemical disinfectant dosages. Biofilm bacteria reacted similarly regardless of the disinfectant dose. Overall, these results indicate that the presence of secondary disinfectants within the distribution system is a crucial component required for microbial control.

Role of NOM on UV Disinfection

The presence of NOM in the source water being pumped into the UV lamp resulted in a similar suspended bacteria population as when NOM was not added. Conversely, NOM was shown to support a biofilm population that was 1-log higher than without any added NOM. For all three disinfectants investigated reductions in heterotrophic bacteria (suspended and biofilm) were less in the presence of NOM than when the reactors were supplemented with BOM. This may have been the result of an increased disinfectant demand from the NOM and the increased concentration of bacteria within the biofilm. NOM was also shown to deposit on both the quartz tube within the UV disinfection unit and in the inner surfaces of the reactor. Oxidation of NOM with UV and a combination of UV and chlorine was shown to increase bioavailability and the production of organic by-products, particularly HAAs and aldehydes. Overall, these results suggest that the amount of NOM in the source water entering the plant should be limited, through pre-treatment steps, in order adequately irradiate the water and control biological growth

Chlorination applied prior to UV disinfection, in the presence of NOM, acted to inhibit the development of biofilms and control the presence of bacteria in the bulk water. Pre-chlorination was able to limit the presence of bacteria to the same extent as was evident when secondary disinfection was applied instead. The presence of chlorine during UV disinfection was also able to control the amount of organic material deposited on the quartz tube as well as on the internal surfaces of the reactors. Overall, these findings indicate that the use of chlorination prior to UV disinfection improves microbial control in distribution systems, and enhances the effectiveness of secondary disinfectants.

RECOMMENDATIONS TO THE WATER INDUSTRY

Practical Implications of Research

UV Calibration

Current UV calibration techniques employ collimated beams to determine the fluence supplied by a particular UV lamp. Although this technique can be an effective and efficient means of determining applied fluence it has limitations. Collimated beams are unable to properly account for lamp fouling, water quality and hydraulic conditions that exist within the reactor at the time of treatment. Calibration of UV disinfection system should be completed on site with water quality and hydraulic conditions that will exist during operation.

UV Treatment Integration

This research considered the application of UV disinfection at various locations during water treatment and distribution. Figure 9.1 illustrates the locations that were represented during Phase 1 (P1), Phase 2 (P2) and Phase 3 (P3) of the research.

In areas where advanced water treatment exists the addition of UV disinfection would enhance existing treatment barriers (e.g., filtration) for microbiological control. Results from Phase 1 showed that UV disinfection enhanced microbial control in distribution systems. If UV disinfection is considered to be a viable treatment option, the utility must decide the optimal location for disinfection in the process train. Following filtration, for example, the water is of very high quality, low turbidity and low colour; consequently, substantial UV disinfection would be possible with a minimal amount of quartz sleeve fouling. Alternatively, if UV light is used prior to filtration, the water is of lower quality and depending on water quality, turbidity and colour, microbial inactivation can be reduced and quartz sleeve fouling may become an issue. In either case it is important to have a UV disinfection system that has been properly calibrated to the water quality and hydraulic conditions to ensure optimum disinfection capacity. Changes in water quality and disinfection efficiency should also be monitored on a regular basis to maintain an adequate level of microbial protection.

In areas where water treatment is nonexistent or minimal, which is the case for most private groundwater wells and some smaller municipal water sources, the addition of UV disinfection could be advantageous. Although numerous groundwater sources have been in use for many years without any perceptible bacteriological problems the addition of UV light would provide a constant barrier against potentially pathogenic bacteria that may arise in the future. Results from Phase 2 indicated that humic rich source water supported an increased population of biofilm bacteria. This may have been the result of UV-oxidation of the humic material. Although UV disinfection may offer an initial microbiological barrier it cannot control the regrowth of biofilm bacteria and secondary disinfection should be employed, particularly in locations where it was used in the past. In this scenario secondary disinfection is applied for additional microbiological control but may result in an increased production of disinfection by-

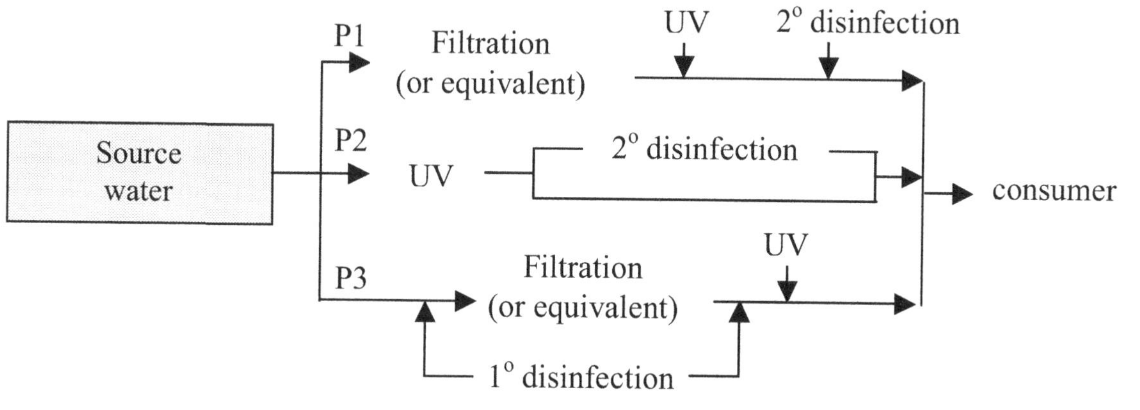

Figure 9.1 Options for integrating UV disinfection into full-scale treatment for minimizing impacts to the distribution system

products during distribution. The operator should be aware of the limitations of UV disinfection. Disinfecting untreated groundwater sources exposes the quartz sleeve to organic and inorganic particles that can cause fouling, which cause the lamp to be less effective. The fluence supplied by the lamp should be checked on a routine basis to ensure adequate disinfection.

In some cases chemical disinfection may be used prior to UV disinfection, as was the case in Phase 2 and Phase 3 of the research. Primary disinfection is used to retard microbial growth during treatment but can also act to form disinfection by-products. The implementation of UV disinfection was shown to increase the production of aldehydes and HAAs. Phase 3 considered the effect of oxidation rather than the biocidal effects of chlorine, which were examined in Phase 2. In practice, these biocidal effects would have a significant effect on growth and would be enhanced by UV disinfection as shown in Phase 1. UV and UV/Cl_2 oxidation increased biological degradation and, although not discussed previously, could be employed as a pre-treatment step to biological filtration, as is the case with ozone, in order to remove organics prior to additional treatment and chemical disinfection.

Identification of Future Research Needs

This study evaluated the impact that UV disinfection has on biological stability in drinking water distribution systems. Results from this study have lead to the identification of future research that is required in order to better understand the impact of UV light on various aspects of drinking water treatment.

Data is needed to determine the impact that various pre-chlorination concentrations have on UV disinfection. Experimentation during this study focused on only one pre-chlorination concentration. This gave an on/off type of comparison. Additional pre-chlorination doses should be investigated to determine if lower concentrations are adequate to control regrowth events. The effect of pre-chlorination and UV disinfection in the absence of humic material should also be investigated for comparison purposes.

The potential to produce UV disinfection by-products should be investigated further. Various by-products are known to form after oxidation with UV light. In this study, the concentration of HAAs and aldehydes both increased with increased oxidation. The impact of increased oxidation and oxidation type should be considered to determine what level of oxidation produces the minimum of DBPs with an adequate level of disinfection, as well as the mechanisms behind the production.

A potential synergy between UV light and chemical disinfection was also noted in some treatments. Research is needed to further examine this synergistic effect, especially with specific pathogenic microorganisms.

The relevance of using representative biological organic matter for nutrients in AR studies is of concern. In this study, humic material was shown to support a denser biofilm with higher total and heterotrophic bacterial cells than the synthetic cocktail. The synthetic cocktail represents nutrients (carbon, nitrogen, and phosphorous) that are found in natural water sources and are easily biodegraded, whereas the humic material represents natural water with turbidity, colour, and organic carbon that are not as easily utilized by bacteria. It is unclear at this time which of the two nutrient sources more closely models natural water conditions, and which should be used for further bench scale studies.

REFERENCES

Agbekodo, M.K., B. Legube, and P. Coté. 1996. Organics in NF Permeate. *Jour. AWWA*, 88(5):67-74.

Allard, B., H. Borén, C. Pettersson, and G. Zhang. 1994. Degradation of Humic Substances by UV Irradiation. *Environment International*, 20(1):97-101.

Anderson, W.B., P.M. Huck, R.M. Slawson, and A.K. Camper. 1997. BOM Component Evolution During Drinking Water Treatment and Distribution. In *Proc. of the AWWA Annual Conference and Exposition*. Atlanta, Georgia.: AWWA

APHA, AWWA, and WEF (American Public Health Association, American Water Works Association, Water Environment Federation). 1995. *Standard Methods for the Examination of Water and Wastewater*. 19th ed. Washington D.C.: APHA.

AWWA (American Water Works Association) and ASCE (American Society of Civil Engineers). 1998. *Water Treatment Plant Design*, 3rd ed. New York: McGraw-Hill.

AWWA Water Quality Division Disinfection Systems Committee. 2000. Disinfection at Large and Medium-size Systems. *Jour. AWWA*, 92(5):32-43.

Backlund, P. 1992. Degradation of Aquatic Humic Material by Ultraviolet Light. *Chemosphere*, 25(12):1869-1878.

Bailey, J.E., and Ollis, D.F. 1986. *Biochemical Engineering Fundamentals* 2nd ed. New York: McGraw-Hill.

Battigelli, D.A., Sobsey, M D., and Lobe, D.C. 1993. The Inactivation of Hepatitis A Virus and Other Model Viruses by UV Irradiation. *Water Science & Technology*, 27(3-4):39-342.

Bitton, G., Y. Henis, and N. Lahav. 1972. Effect of Several Clay Minerals on the Survival of *Klebsiella aerogenes* Exposed to Ultraviolet Irradiation. *Applied Microbiology*, 23(5):870-874.

Blatchley, E.R., R.W. Emerick, T. Hargy, O. Hoyer, R.W. Hultquist, R.H. Sakaji, O.K. Scheible, D.C. Schmelling, F. Soroushian, and G. Tchobanogolous. 2000. *Ultraviolet Disinfection: Guidelines for Drinking Water and Water Reuse*, NWRI and AwwaRF, Fountain Valley, Calif.

Block, J.C. 1995. Survival and Growth of E. coli in Drinking Water Distribution Systems. *Proceedings of the International Conference on Coliforms and E. coli*. Leeds, UK.

Block, J.C., L. Mathieu, P. Servais, D. Fontvieille, and P. Werner. 1992. Indigenous Bacterial Inocula for Measuring the Biodegradable Dissolved Organic Carbon (BDOC) in Waters. *Water Research*, 26(2):481-486.

Bryers, J.D. 1987. Biologically Active Surfaces: Processes Governing the Formation and Persistence of Biofilms. *Biotechnology Progress*, 3(2):57-68.

Bukhari, Z., T.M. Hargy, J.R. Bolton, B. Dussert, and J.L. Clancy. 1999. Medium-pressure UV for Oocyst Inactivation. *Jour. AWWA*, 91(3):86-94.

Camel, V., and A. Bermond. 1998. The Use of Ozone and Associated Oxidation Processes in Drinking Water Treatment. *Water Research*, 32(11):3208-3222.

Camper, A.K. 1996. *Factors Limiting Microbial Growth in Distribution Systems: Laboratory and Pilot-Scale Experiments*. Denver, Colo.: AwwaRF and AWWA.

Camper, A.K. 1995. Factors Influencing Biofilm Growth in Drinking Water Distribution Systems. Ph.D. diss., Montana State University-Bozeman, Bozeman, MT.

Camper, A.K., J. Buls, and L. Goodrum. 2001. Effects of UV Disinfection on Humic Substances and Biofilms. In *Proc. of the 2001 Annual AWWA Water Quality Technology Conference*. Nashville, TN.: AWWA.

Camper, A.K., L. Goodrum, and J. Buls. 2001. *Use of Ultraviolet Light for Disinfection of Drinking Water*. Final Report to the Drinking Water Assistance Program for Small Systems, Montana State University, Bozeman, MT.

Camper, A.K., W.L. Jones, and J.T. Hayes. 1996. Effects of Growth Conditions and Substratum Composition on the Presence of Coliforms in Mixed-population Biofilms. *Journal of Environmental Microbiology*, 62(11):4014-4018.

Chandy, J.P., and M.L. Angles. 2001. Determination of Nutrients Limiting Biofilm Formation and the Subsequent Impact on Disinfection Decay. *Water Research*, 35(11): 2677-2682.

Chauret, C., R. Creason, J. Jarosh, C. Volk, J. Robinson, and C. Warnes. 2001. Detection of *Aeromonas hydrophila* in a Drinking Water Distribution System: a Field and Pilot Study. *Canadian Journal Microbiology*, 47:782-786.

Chen, W.J., and C.P. Weisel. 1998. Halogenated DBP Concentrations in a Distribution System. *Jour. AWWA*, 90(4):151-163.

Chiswell, B., and K.R. O'Halloran. 1991. Use of Lissamine Green B as a Spectrophotometric Reagent for the Determination of Low Residuals of Chlorine Dioxide. *Analyst*, 116:657-661.

Cipparone, L.A., A.C. Diehl, and G.E. Speitel Jr. 1997. Ozonation and BDOC Removal: Effect on Water Quality. *Jour. AWWA*, 89(2):84-97

Corin, N., P. Backlund, and M. Kulovaara. 1996. Degradation Products Formed During UV-Irradiation of Humic Waters. *Chemosphere*, 33(2):245-255.

Cotton, C.A., D.M. Owen, G.C. Cline, and T.P. Brodeur. 2001. UV Disinfection Costs for Inactivating *Cryptosporidium*. *Jour. AWWA*, 93(6):82-94.

Craik, S.A., D. Weldon, G.R. Finch, J.R. Bolton, and M. Belosevic. 2001. Inactivation of *Cryptosporidium parvum* oocysts using Medium- and Low-pressure Ultraviolet Radiation. *Water Research*, 35(6):1387-1398.

Craik, S.A., G.R. Finch, J.R. Bolton, and M. Belosevic. 2000. Inactivation of *Giardia muris* Cysts Using Medium Pressure Ultraviolet Radiation in Filtered Drinking Water. *Water Research*, 34(18):4325-4332

Craun, G.F., S.A. Hubbs, F. Frost, R.L. Calderon, and S.H. Via. 1998. Waterborne Outbreaks of Cryptosporidiosis. *Jour. AWWA*, 90(9):81-91.

Droste, R.L. 1997. *Theory and Practice of Water and Wastewater Treatment*. New York.: John Wiley & Sons.

Ellis, B.D., P. Butterfield, W.L. Jones, G.A. McFeters, and A.K. Camper. 2000. Effects of Carbon Source, Carbon Concentration, and Chlorination on Growth Related Parameters of Heterotrophic Biofilm Bacteria. *Microbial Ecology*, 38:330-347.

Dykstra, T.S., C. Chauret, and G.A. Gagnon. 2002. Hydraulic Calibration and Fluence Determination of a Model UV Disinfection System. *ASCE – J. Enviro. Engrg*. In press November 2002.

Dykstra, T.S., K.C. O'Leary, C. Chauret, R.C. Andrews, and G.A. Gagnon. Impact of UV Disinfection on Biofilm Growth in Distribution Systems. *2002 AWWA Water Quality Technology Conference Proceedings.* Denver, Colo: AWWA.

Escobar, I.C., and A.A. Randall. 2001. Case Study: Ozonation and Distribution System Biostability. *Jour. AWWA*, 93(10):77-89.

Frías, J., F. Ribas, and F. Lucena. 1992. A Method for the Measurement of Biodegradable Organic-carbon in Waters. *Water Research*, 26(2):255-258.

Gagnon, G.A. 1997. *Utilization of Biodegradable Organic Matter by Biofilms in Drinking Water Distribution Systems.* Ph.D. diss., University of Waterloo, Waterloo, ON.

Gagnon, G.A., and R.M. Slawson. 1999. An Efficient Biofilm Removal Method for Bacterial Cells Exposed to Drinking Water. *Journal of Microbiological Methods*, 34:203-214.

Gagnon, G.A., P.M. Huck, G.S. Irwin, and P.J. Ollos. 1998. Defining the Relationships Between BOM, Chlorine Residual and Bacterial Regrowth. *Proceedings, IWSA/AWWA Joint Symposium on Protecting Water Quality in the Distribution System: What is the Role of Disinfectant Residuals.* Philadelphia, PA.

Gatel, D., Dernat, M., and Bazer-Bachi, A. 1995. Drinking Water use of Chlorine Dioxide: A Case Study in France. *Proceedings of the Third International Symposium on Chlorine Dioxide.* CMA, AwwaRF and U.S. EPA. New Orleans, LA

Gauthier, V., B. Gérard, J-M. Portal, J-C. Block, and D. Gatel. 1999. Organic Matter as Loose Deposits in a Drinking Water Distribution System. *Water Research*, 33(4):1014-1026.

Geldreich, E.E. 1996. *Microbial Quality of Water Supply in Distribution Systems.* Boca Raton, FL.: CRC Press.

Gjaltema, A., P.A.M. Arts, M.C.M. van Loosdrecht, J.G. Kuenen, and J.J. Heijnen. 1994. Heterogeneity of Biofilms in Rotating Annular Reactors: Occurrence, Structure, and Consequences. *Biotechnology Bioengineering*, 44:194-204.

Gjessing, E.T., and T. Källqvist. 1991. Algicidal and Chemical Effect of U.V.-Radiation of Water Containing Humic Substances. *Water Research*, 25(4):491-494.

Goel, S., R.M. Hozalski, and E.J. Bouwer. 1995. Biodegradation of NOM: Effect of NOM Source and Ozone Dose. *Jour. AWWA*, 87(1):90-105.

Graham, N.J.D. 1999. Removal of Humic Substances by Oxidation/biofiltration Processes – A Review. *Water Science & Technology*, 40(9):141-148.

Haas, C.N. 1999. Benefits of Using a Disinfectant Residual. *Jour. AWWA*, 91(1):65-69.

Hanninen, M.L., S. Salmi, L. Mattila, R. Taipalinen, and A. Siitonen. 1995. Association of *Aeromonas* spp. with Travellers' Diarrhoea in Finland. *Journal Medical Microbiology*, 42:26-31.

Harris, G.D., Adams, D., Moore, W. M., and Sorrensen, D.L. 1987. Potassium Ferrioxalate as Chemical Actinometer in Ultraviolet Reactors. *Journal of Environmental Engineering*, ASCE, 113(3):612-627.

Havelaar, A.H., C.C.E. Meulemans, W.M. Pot-Hogeboom, and J. Koster. 1990. Inactivation of Bacteriophage MS2 in Wastewater Effluent with Monochromatic and Polychromatic Ultraviolet Light. *Water Research*, 24(11):1387-1393.

Hozalski, R.M., E.J. Bouwer, and S. Goel. 1999. Removal of Natural Organic Matter (NOM) from Drinking Water Supplies by Ozone-biofiltration. *Water Science & Technology*, 40(9):157-163.

Huck, P.M. 1990. Measurement of Biodegradable Organic Matter and Bacterial Growth Potential in Drinking Water. *Jour. AWWA,* 82(7):78-86.

Huys, G., P. Kampfer, M. Altweeg, I. Kersters, A. Lamb, R. Coopman, J. Luthy-Hottenstein, M. Vancaneyt, P. Janssen, and K. Kersters. 1997. *Aeromonas popoffii* sp. nov., a Mesophilic Bacterium Isolated from Drinking Water Production Plants and Reservoirs. *International Journal Systematic Bacteriology*, 47:1165-1171.

ISO (International Organization for Standardization). 1995. Water Quality – Detection and Enumeration of F-specific RNA bacteriophages, 1, Method by Incubation with a Host Strain. *Rep. ISO* 10705-1, Geneva.

Jagger, J. H. 1967. *Introduction to Research in UV Photobiology*. Englewood Cliffs, NJ.: Prentice-Hall.

Jones, W.L., A.K. Camper, X. Xu, and S. Bakich. 1995. Direct and Indirect Measurements of Microbial Growth in Drinking Water Distribution Systems. In *Proc. of the 1995 Annual AWWA Water Quality Tecchnology Conference*. New Orleans, LA.: AWWA.

Juang, D-F, and J.M. Morgan. 2001. The Applicability of the API 20E and API Rapid NFT Systems for the Identification of Bacteria from Activated Sludge. *Environmental Biotechnology*, 4:1-6.

Kaplan, L.A., D.J. Reasoner, and E.W. Rice. 1994. A Survey of BOM in U.S. Drinking Waters. *Jour. AWWA*, 86(2):121-132.

Kemmy, F.A., J.C. Fry, and R.A. Breach. 1989. Development and Operational Implementation of a Modified and Simplified Method for Determination of Assimilable Organic Carbon (AOC) in Drinking Water. *Water Science & Technology*, 21(3):155-159.

Korich, D.G., J.R. Mead, M.S. Madore, N.A. Sinclair, and C.R. Sterling. 1990. Effects of Ozone, Chlorine Dioxide, Chlorine and Monochloramine on *Cryptosporidium parvum* Oocyst Viability. *Applied Environmental Microbiology*, 56:1423-1428.

Kouame, Y, and C.N. Haas. 1991. Inactivation of *E. coli* by Combined Action of Free Chlorine and Monochloramine. *Water Research*, 25(9):1027-1032.

Krovacek, K., A. Faris, S. Baloda, T. Lindberg, M. Peterz, and I. Månsson, I. 1992. Isolation and Virulence Profiles of *Aeromonas* spp. from Different Municipal Drinking Water Supplies in Sweden. *Food Microbiology*, 9:215-222.

Kühn, I., G. Allestam, G. Huys, P. Janssen, K. Kersters, K. Krovacek, and T-A Stenström. 1997. Diversity, Persistence, and Virulence of *Aeromonas* strains Isolated from Drinking Water Distribution Systems in Sweden. *Applied Environment Microbiology*, 63:2708-2715.

LeChevallier, M.W., C. Norton, A. K. Camper, P. Morin, B. Ellis, W. Jones, A. Rompe, M. Prevost, J. Coallier, P Servais, D. Holt, A. Delanoue, and J. Colbourne. 1998. Microbial Impact of Biological Filtration. Denver, Colo.: AwwaRF and AWWA.

LeChevallier, M.W., N.J. Welch, and D.B. Smith. 1996. Full-scale Studies Related to Colifom Regrowth in Drinking Water. *Applied Environmental Microbiology*, 62:2201-2211.

LeChevallier, M.W., C.D. Lowry, R.G. Lee, and D.L. Gibbon. 1993. Examining the Relationship Between iron corrosion and the disinfection of biofilm bacteria. *Jour. AWWA,* 85(7): 111-123.

LeChevallier, M.W., W. Schulz, and R.G. Lee. 1991. Bacterial Nutrients in Drinking Water. Appl. Environ. Microbial. 57(3):857-862.

LeChevallier, M.W. 1990. Coliform Regrowth in Drinking Water. *Jour. AWWA*, 82(11):74-86.

LeChevallier, M.W., T.M. Babcock, and R.G. Lee. 1987. Examination and Characterization of Distribution System Biofilms. *Applied Environmental Microbiology*, 53(12): 2714-2724.

Levenspiel, O. 1999. *Chemical Reaction Engineering* 3rd ed. New York. John Wiley & Sons.

Limoni, B. and B. Teltsch. 1985. Chlorine Dioxide Disinfection of Drinking Water - An Evaluation of a Treatment Plant. *Water Research*, 19:1489-1495.

Linden, K.G. and A.A. Mofidi. 1999. *Measurement of UV Irradiance: Tools and Considerations.* In *Proc. of the 1999 Annual AWWA Water Quality Technology Conference* Tampa, FL.: AWWA.

Lu, W., L. Kiéné, and Y. Lévi. 1999. Chlorine Demand of Biofilms in Water Distribution Systems. *Water Research*, 33(3):827-835

Lund, V., and D. Hongve. 1994. Ultraviolet Irradiated Water Containing Humic Substances Inhibits Bacterial Metabolism. *Water Research*, 28(5):1111-1116.

Lyn, D.A., K. Chiu, and E.R. Blatchley. 1999. Numerical Modeling of Flow and Disinfection in UV Disinfection Channels. *Journal of Environmental Engineering*, ASCE, 125(1):17-26.

MacCarthy, P., and R.L. Malcolm. 1987. The Nature of Commercial Humic Acids. In *Aquatic Humic Substances: Influence on Fate and Treatment of Pollutants*. Denver Colo.: 193rd National Meeting of the American Chemistry Society.

Mathieu, L., J.C. Block, M. Dutang, J. Maillard, and D. Reasoner. 1993. Control of Biofilm Accumulation in Drinking Water Distribution Systems. *Water Supply*. 11(3,4): 365-376.

Mayack, L.A., R.J. Soracco, E.W. Wilde, and D.H. Pope. 1984. Comparative Effectiveness of Chlorine and Chlorine Dioxide Biocide Regimes for Biofouling Control. *Water Research*, 18:593-599.

McGuire, M.J., N.I. Lieu, and M.S. Pearthree. 1999. Using Chlorite Ion to Control Nitrification. *Jour. AWWA*, 91(10):52-61.

Melin, E.S., and H. Ødegaard. 1999. Biofiltration of Ozonated Humic Water in Expanded Clay Aggregate Filters. *Water Science & Technology*, 40(9):165-172.

Miettinen, I.T., T.Vartiainen, and P.J. Martikainen. 1997. Phosphorus and Bacterial Growth in Drinking Water. *Applied Environmental Microbiology*, 68(8):3242–3245.

Mofidi, A.A., H. Baribeau, P.A. Rochelle, R. De Leon, B.M. Coffey, and J.F. Green. 2001. Disinfection of *Cryptosporidium* with Polychromatic UV Light. *Jour. AWWA*, 93(6):95-109.

Montgomery, J.M., 1985. *Water Treatment Principles and Design*, New York, John Wiley & Sons.

Nobel, P.A., D.L. Clark, and B.H. Olson. 1996. Biological Stability of Groundwater. *Jour. AWWA*, 88(5):87-96.

Norton, C.D. and M.W. LeChevallier, 2000. A Pilot Study of Bacteriological Population Changes Through Potable Water Treatment and Distribution. *Appl. Environ. Microbial*. 66(1):268-276.

Norton, C.D. and M.W. LeChevallier. 1997. Chloramination: Its Effect on Distribution System Water Quality. *Jour. AWWA*, 89(7):66-77.

Nowell, L.H., and J. Hoihné. 1992a. Photolysis of Aqueous Chlorine at Sunlight and Ultraviolet Wavelengths – I. Degradation Rates. *Water Research*, 26(5):593-598.

Nowell, L.H., and J. Hoihné. 1992b. Photolysis of Aqueous Chlorine at Sunlight and Ultraviolet Wavelengths – I. Hydroxyl Radical Production. *Water Research*, 26(5):599-605.

Oliveira, R., L.F. Melo, A. Oliveira, and R. Salgueiro. 1994. Polyssacharide Production and Biofilm Formation by *Pseudomonas fluorescens* : Effects of pH and Surface Material. *Colloids and Surfaces B: Biointerfaces*, 2, 41-46.

Payment, P., J. Siemiatycki, L. Richardson, G. Renaud, E. Franco, and M. Prévost. 1997. A Prospective Epidemiological Study of Gastrointestinal Health Effects due to the Consumption of Drinking Water. *International Journal of Environmental Health Research*. 7:5-31.

Pedersen, K. 1990. Biofilm Development on Stainless Steel and PVC Surfaces in Drinking Water. *Water Research*, 24(2):239-243.

Peldszus, S., P.M. Huck and S.A. Andrews. 1998. Quantitative Determination of Oxalate and Other Organic Acids in Drinking Water at Low μg/L Concentrations. *Journal of Chromatogaphy A*, 792:198-203.

Peldszus, S., P.M. Huck, and S.A. Andrews. .1996. Determination of Short Chain aliphatic, oxo- and hydroxy- acids in Drinking Water at low μg/L Concentrations. *Journal of Chromatogaphy A*, 723:27-34.

Percival, S.L., J.T. Walker, and P.R. Hunter. 2000. *Microbiological Aspects of Biofilms and Drinking Water*, Boca Raton, FL.: CRC Press.

Peyton, B.M., and W.G. Characklis. 1993. A Statistical Analysis of the Effect of Substrate Utilization and Shear Stress on the Kinetics of Biofilm Detachment. *Biotechnology and Bioengineering*, 41(7):728-735.

Prévost, M., A. Rompré, H. Baribeau, J. Coallier, and P. Lafrance. 1997. Service Lines: Their Effect on Microbiological Quality. *Jour. AWWA*, 89(7):78-91.

Prévost, M., A. Rompré, J. Coallier, P. Servais, P. Laurent, B. Clément, and P. Lafrance. 1998. Suspended Bacterial Biomass and Activity in Full-scale Drinking Water Distribution Systems: Impact of Water Treatment. *Water Research* 32(5):1393-1406.

Qualls, R.G., M.P. Flynn, and J.D. Johnson. 1983. The Role of Suspended Particles in Ultraviolet Disinfection. *Journal of Water Pollution Control Federation,* 55(10): 1280-1285.

Reckhow, D.A., J.E. Tobiason, W. Pouvesle, J.N. McClellan, J.K. Edzwald, and D. Smith. 1996. Evolution of Natural Organic Matter Through a Drinking Water Distribution System. In *Proc. Influence of Natural Organic Matter Characteristics on Drinking Water Treatment and Quality*. Poitiers, France.

Rittmann, B.E. and V.L. Snoeyink. 1984. Achieving Biologically Stable Drinking Water. *Jour. AWWA*, 76(10):106-114.

Rittmann, B.E., and P.M. Huck. 1989. Biological Treatment of Public Water Supplies. *CRC Critical Reviews in Environment Control*. Boca Raton, FL.: CRC Press.

Sclimentii, M.J., S.W. Krasner, W.H. Glaze, and H.S. Weinberg. 1990. Ozone Disinfection By-products: Optimization of the PFBHA Derivatization Method for the Analysis of Aldehydes. In *Proc. of the 1990 Annual AWWA Water Quality Tecchnology Conference*, San Diego, CA.: AWWA.

Servais, P., A. Anzil, and C. Ventresque. 1989. Simple Method for Determining of Biodegradable Dissolved Organic Carbon in Water. *Applied Environmental Microbiology*, 55(10):2732-2734.

Sharp, R.R., A.K. Camper, J.J. Crippen, O.D. Schneider, and S. Leggiero. 2001. Evaluation of Drinking Water Biostability Using Biofilm Methods. *Journal of Environmental Engineering*, ASCE, 127(5):403-410.

Shaw, J.P., J.P. Malley Jr., and S.A. Willoughby. 2000. Effects of UV Irradiation on Organic Matter. *Jour. AWWA,* 92(4):157-167.

Sheriff, M., and R. Gehr. 2001. Laboratory Investigation of Inorganic Fouling of Low Pressure UV Disinfectioon Lamps. *Canadian Journal of Water Quality Research*, 36(1): 71-92.

Sibille, I., L. Mathieu, J.L. Paquin, D. Gatel, and J.C. Block. 1997. Microbial Characteristics of a Distribution System fed with Nanofiltered Drinking Water. *Water Research*, 31: 2318-2326.

Snicer, G.A., J.P. Malley, A.B. Magolin, and S.P. Hogan. 1996. Evaluation of Ultraviolet Technology in Drinking Water Treatment. In *Proc. of the 1996 Annual AWWA Water Quality Tecchnology Conference*, Boston, MA.: AWWA.

Speitel, G.E., and F.A. DiGiano. 1987. Biofilm Shearing Under Dynamic Conditions. *Journal of Environmental Engineering*, ASCE, 113:464-475.

Stewart, P.S. 1993. A Model of Biofilm Detachment, *Biotechnology and Bioengineering*, 41(1):111-117.

Storey, M.V., and N.J. Ashbolt. 2001. Persistence of two Model Enteric Viruses (B40-8 and MS-2 Bacteriophages) in Water Distribution Pipe Biofilms. *Water Science & Technology*, 43(12):133-136.

Suffet, I.H., A. Corado, D. Chou, M.J. McGuire, and S. Butterworth. 1996. AWWA Taste and Odor Survey. *Jour. AWWA*, 88(4):168-180.

Thomas, J.M., W.A. McKay, E. Cole, J.E. Landmeyer, and P.M. Bradley. 2000. The Fate of Haloacetic Acids and Trihalomethanes in an Aquifer Storage and Recovery Program, Las Vegas, Nevada. *Ground Water*, 38(4):605-614.

Trussel, R.R. 1999. Safeguarding Distribution System Integrity. *Jour. AWWA*, 91(1):46-54.

van der Kooij, D., A. Visser, and W.A.M. Hijnen. 1982. Determining the Concentration of Easily Assimilable Organic Carbon in Drinking Water. *Jour. AWWA*, 74(10):540-545.

van der Kooij, D., J.H.M. van Lieverloo, J. Schellart, and P. Hiemstra. 1999. Maintaining quality without a disinfectant residual. *Jour. AWWA*, 91(1):55-64.

van der Wende, E., W.G. Characklis, and D. B. Smith. 1989. Biofilms and Bacterial Drinking Water Quality. *Water Research*, 23(10):1313-1322.

Vieira, M.J., and L.F.Melo. 1995. Effect of Clay Particles on the Behaviour of Biofilms Formed by *Pseudomonas fluorescens*. *Water Science and Technology*, 32 (8), 45-52.

Vallino, J.J., C.S. Hopkinson, and J.E. Hobbie. 1996. Modeling Bacterial Utilization of Dissolved Organic Matter: Optimization Replaces Monod Growth Kinetics. Liminology and Oceanography, 41(8):1591-1609.

Volk C., and M.W. LeChevallier 1999. Impacts of the Reduction of Nutrient Levels on Bacterial Water Quality in Distribution systems. *Appl. Environ. Microbiol.*, 65: 4957-4966.

Volk, C.J., R. Hoffman, C. Chauret, G.A. Gagnon, G. Rangers and R.C. Andrews. 2002. Implementation of Chlorine Dioxide Disinfection: Effects of the Treatment Change on Drinking Water Quality in a Full-scale Distribution System. *Journal of Environmental Engineering and Science*, forthcoming.

Weber, W.J. Jr., and F.A. DiGiano. 1995. *Process Dynamics in Environmental Systems*. New York, John Wiley & Sons.

Weinberg, H.S., W.H. Glaze, S.W. Krasner, and M.J. Sclimenti. 1993. Formation and Removal of Aldehydes in Plants that use Ozonation. *Jour. AWWA*, 85(5):72-85.

Wilczak, A., J.G. Jacangelo, J.P. Marcinko, L.H. Odell, G.J. Kirmeyer, and R.L. Wolfe. 1996. Occurrence of Nitrification in Chloraminated Distribution Systems, *Jour. AWWA*, 88(7):74-85.

Zacheus, O.M., M.J. Lehtola, L.K. Korhonen, and P.J. Martikainen (2001), Soft Deposits, the Key Site for Microbial Growth in Drinking Water Distribution System Networks, *Water Research*, 35(7) 1757-1765

Zheng, M., S.A. Andrews, and J.R. Bolton. 1999. Impacts of Medium-Pressure UV on THM and HAA Formation in Pre-UV Chlorinated Drinking Water, In *Proc. of the 1999 AWWA Water Quality Technology Conference*. Denver, Colo.: AWWA.

ABBREVIATIONS

AODC	Acridine Orange Direct Count
AR	Annular Reactor
BAC	Biologically Activated Carbon
BCAA	Bromodichloroactic acid
BOM	Biodegradable Organic Matter
C	concentration (M)
CFU	Colony Forming Units
D	dispersion (cm^2/s)
DBAA	Dibromoacetic acid
DBP	Disinfection by-product
DCAA	Dichloroacetic acid
DOC	Dissolved Organic Carbon
DOM	Dissolved Organic Matter
EBMUD	East Bay Municipal Utilities District
GAC	Granular Activated Carbon
HAA	Haloacetic Acid
HPC	Heterotrophic Plate Counts
HRT	Hydraulic Retention Time
HRWC	Halifax Regional Water Commission
ISO	International Organization for Standardization
k	decay coefficient (s^{-1})

L	length of reactor (cm)
m	mass flow rate (kg/s)
MCAA	Monochloroacetic acid
N	number of bacteria surviving after UV exposure
N_o	initial number of bacteria
NOM	Natural Organic Matter
NPOC	Nonpurgeable Organic Carbon
NWRI	National Water Research Institute
PBS	Phosphate Buffered Saline
pfu	Plaque forming units
RTD	Residence Time Distribution
s_3	skewness (min^{-1})
t	time (s)
TCAA	Trichloroacetic acid
THM	Trihalomethane
t_m	mean residence time (s)
TOC	Total Organic Carbon
u	fluid flow (cm/s)
UV	Ultraviolet
x	photolytic reaction length (cm)
Φ	quantum yield (moles Fe^{2+}/einstein)
ν	fluid flowing through a vessel of volume V (m^3/s)
$\sigma^2_{\theta^o}$	dimensionless form of variance

IWA Publishing's authorised EU representative for General Product Safety Regulations is Diane D'Arras, 15 rue Duret, 75116 Paris, France, e-mail: safety@iwap.co.uk.

Printed and bound by CPI Group (UK) Ltd, Croydon, CR0 4YY
17/09/2026
02218813-0017